Separations Division

Core Programming Topic at the 2012 AIChE Spring Meeting and 8th Global Congress on Process Safety

Houston, Texas, USA
1-5 April 2012

ISBN: 978-1-62276-553-9

Printed from e-media with permission by:

Curran Associates, Inc.
57 Morehouse Lane
Red Hook, NY 12571

Some format issues inherent in the e-media version may also appear in this print version.

Printed by Curran Associates, Inc. (2012)

For permission requests, please contact AIChE
at the address below.

AIChE
3 Park Avenue
New York, NY 10016-5991

Phone: (203) 702-7660
Fax: (203) 775-5177

www.aiche.org

Additional copies of this publication are available from:

Curran Associates, Inc.
57 Morehouse Lane
Red Hook, NY 12571 USA
Phone: 845-758-0400
Fax: 845-758-2634
Email: curran@proceedings.com
Web: www.proceedings.com

TABLE OF CONTENTS

(27a) Development of Structure/Property Correlations for Water and Ion Transport in Polymers

Dr. Benny D. Freeman, Chemical Engineering, University of Texas, Austin, TX

The development of advanced polymer membranes for water purification, in applications such as desalination via reverse osmosis, forward osmosis, nanofiltration, membrane-assisted capacitive deionization, or electrodialysis, and in energy generation applications such as forward osmosis and reverse electrodialysis, relies on an intimate understanding of the fundamental relationship between polymer structure and ion and water transport properties. Much of this understanding does not currently exist. This presentation will describe efforts to characterize ion and water sorption, diffusion and permeation in several systematic series of polymers. We find strong correlations between free volume in hydrated polymers, as characterized via water uptake or results from positron annihilation lifetime spectroscopy studies, and ion and water diffusion and permeation in polymers. We also find that there is a distinct tradeoff between water permeability (and diffusivity) and water/ion permeability (and diffusivity) selectivity. Efforts to independently control membrane charge and water uptake can yield polymers with improved transport and separation properties.

(27b) Engineered Osmosis: A Rapidly Emerging Separation Technology Platform

Jeffrey R. McCutcheon, Department of Chemical, Materials and Biomolecular Engineering, University of Connecticut, Storrs, CT

The shortage of energy and water resources is a grand challenge facing humanity in the 21stcentury. Engineered osmosis (EO) technologies have been increasingly investigated as an alternative means of sustainable water and energy production. These membrane technologies exploit the gradient of osmotic pressure between a dilute feed solution and a concentrated draw solution as a driving force for water treatment (forward osmosis) and electricity generation (pressure retarded osmosis). This tutorial will provide an overview of this rapidly emerging membrane separations area. Among the highlighted topics will be a detailed description of each EO process, promised benefits through replacement of or hybridization with existing separation processes, and the obstacles to successful development. This talk will also consist of a general overview of how these obstacles are being addressed in academic and industrial groups around the world.

(27c) Functionalization of Polybenzimidazole Membranes for Pore Size Reduction, Increased Hydrophilicity and Surface Charge

Michael Flanagan[1], Richard Hausman[2], Isabel C. Escobar[3], Maria R. Coleman[1] and Tai Shung Chung[4], (1)Department of Chemical and Environmental Engineering, University of Toledo, Toledo, OH, (2)The University of Toledo, Toledo, OH, (3)Chemical and Environmental Engineering, The University of Toledo, Toledo, OH, (4)Department of Chemical and Biomolecular Engineering, National University of Singapore, Singapore, Singapore

Forward osmosis involves the movement of water across a selectively permeable membrane. Unlike pressure driven processes, such as reverse osmosis (RO), the driving force for water permeation through the membrane in FO is the difference in osmotic pressure between the feed and draw solutions. Recently, a hollow fiber nanofiltration (NF) forward osmosis membrane made of polybenzimidazole (PBI) was developed by researchers in Singapore. PBI is a material with outstanding chemical resistance and thermal and mechanical stability. PBI is a promising material for FO membrane development due to its ability to generate high water flux in FO processes. The drawbacks with PBI include low salt rejection in the virgin state and hydrophobicity.

This study focused on functionalization of PBI membrane surfaces using several different modifying agents, with the goal of reducing pore size, increasing hydrophilicity and imparting a surface charge. The surface charge was expected to yield an increased rejection of ions and charged particles in the feed solution. The modifying agents chosen for functionalization included: taurine, ethylene diamine, para-phenylene diamine and poly(acrylamide-co-acrylic acid). These modifying agents were selected for their potential to increase hydrophilicity and impart a charge on the PBI surface, when in near neutral pH environments.

Membranes were characterized using several analytical techniques to verify that modification was successful. Fourier transform infrared spectroscopy in attenuated reflectance mode (FTIR-ATR) was used to detect changes associated with each of the modifying agents. Zeta potential measurements were taken to detect changes in the membranes surface charge. Environmental scanning electron microscopy (ESEM) was used to detect any changes in morphology of the membranes during and after modification. Contact angle measurements were taken to determine changes in hydrophilicity. Lastly, chemical force microscopy (CFM) analysis was used to compare changes in hydrophilicity and charge characteristics. Functionalization, surface charge, and increased hydrophilicity were all verified. In addition, transport characteristics were evaluated for all membrane modifications and virgin samples.

(27d) Industrial Test Trials with Membrane Distillation

Andrea Metz, Zimpro, Siemens , Rothschild, WI,**Clayton B. Maugans**, Zimpro, Siemens, Rothschild, WI and Chan Tun, Siemens Water Technologies, South Windsor, Australia

This paper will summarize the potential applications of membrane distillation process such as seawater desalination and water reclamation from industrial effluent. Performance of MD process will be presented from numerous synthetic and actual feeds. A spectrum of membrane distillation configurations was also evaluated.

(27e) Pervaporative Separation of Water-Alcohol Mixtures Using Ghanaian Zeolite-Clay Filled PVA Composite Membranes

Emmanuel G. Ankudeyand Mohammed N Zainudeen, Chemical Engineering, Kwame Nkrumah University of Science and Technology, Kumasi, Ghana

A pervaporative separation of ethanol-water mixtures was successfully performed using a Polyvinyl alcohol (PVA) based composite membrane filled with zeolites made from local clays from Ghana. The substrate for the ultra-thin PVA composite membrane was made of polysulphone. The pervaporation cell was fabricated locally from scrap aluminium and feed concentrations of ethanol-water mixtures ranging from 60 to 90 v% were separated.

The clay samples used in the development of the membranes were obtained from three different locations of Abonku, Anfoega and Teleku-Bukoazu. The effective surface area of the membranes for separation was 0.007854 m^2. All separations were performed at room temperature and the feed was at atmospheric pressure.

For all the feed concentrations, the composite membrane with Teleku-Bukoazu zeolite-clay filler produced the best performance after ten hours. For a feed concentration of 87.5 wt %, the mixture was enriched to concentrations more than 95 wt% and a separation factor of 85.1 g/m^2-h was obtained.

(30a) AIChE Column Performance Evaluation Guidelines

Ping Xu, UOP LLC, Tonawanda, NY, Daniel R. Summers, Sulzer Chemtech USA, Inc., Tulsa, OK and T.J. Cai, Fractionation Research, Inc., Stillwater, OK

The AIChE has a committee that addresses regularly AIChE's equipment testing procedures. That committee includes a subcommittee that addresses the testing of trayed and packed columns. After three years of work, that subcommittee is ready to release a 150-page manual entitled, "Trayed and Packed Columns – A Guide to Performance Evaluation." This will be the third edition of this manual. This presentation describes the third edition, especially changes with respect to earlier editions.

(30b) Modeling Solvents for Amine-Based CO2 Absorption

Prakash Karpe, Clint P. Aichele and David H. Van Wagener, Sustainability Technologies, ConocoPhillips Company, Bartlesville, OK

Aqueous amine absorption is a mature and proven technology for treating natural gas and refinery fuel gas containing $CO_2$and H_2S. For that reason, it will be a key technology for capturing $CO_2$from flue gas streams for carbon capture and storage. However, the large gas volume and low pressure associated with flue gas streams will result in process units much larger and more expensive than conventional acid gas treaters. Optimization of the design of the absorber and regenerator is important for full-scale CO_2capture in order to minimize both capital and operating costs. This presentation focuses on the development of process modeling using rate-based modeling software for amine absorption. The modeling conclusions incorporate appropriate factors to accurately calculate the overall CO_2avoided, and proper technology comparison strategies are discussed. Regeneration energy comparisons are made between the conventional, 30 wt.% MEA system and other solvents of interest.

(30c) Rate Based Modeling - the Developer's Perspective

Nathan A. Hatcher, Optimized Gas Treating, Inc., Buda, TX

20^{+}years of research, development, and commercialization have culminated in the wide acceptance and success of the ProTreat mass transfer rate-based gas treating simulator. As engineers, we tend to focus on the end product without giving much thought to the thoughts and decisions that are required along the journey to reach a goal. A glimpse into the mindset of engineering software development from the developer's perspective is provided. In particular, certain traits are almost pre-requisites for success in this area of chemical engineering.

(30d) Comparison of Rate-based Results Vs Nutter Ring FRI Test Results using ChemSep

Attilio J. Praderio, LNG Technology Development, ConocoPhillips, Houston, TN and Ross Taylor, Clarkson University, Potsdam, NY

ChemSep is a stand-alone single column simulator for distillation, absorption, and extraction operations. It combines the classic equilibrium stage column model with a nonequilibrium (rate-based) column model in a single program. In this work ChemSep was used to simulate tests carried out by FRI on Nutter Rings #2 (selected from FRI Topical Report # 99) in order to compare the results from several rate base simulations against FRI reported field results. Simulated results matched field results but, as with any simulation software, different results can be obtained when inappropriate options are selected for the specific case under study. The effect of the choice of the binary interaction parameters for the selected equation of state, the choice of the mass transfer models, and methods used for estimation of physical properties of the mixtures are demonstrated in the presentation.

(30e) New Strategies for Acid Gas Enrichment

Tofik K. Khanmamedov, TKK COMPANY, Sugar Land, TX

Sulphur plant performance and capacity are greatly affected by feed gas quality. Thus, increasing the H_2S content of Claus plant feeds is needed when plant throughput is limited or the sour gas has too low an $H_2S:CO_2$ratio for producing satisfactory acid-gas. The key to acid gas enrichment (AGE) is selectivity, and of the generic amines,*N*-methyldiethanolamine (MDEA) is the most selective. The FLEXSORB® family of hindered amines is even more selective; however, these solvent are more expensive than generics. Recently, a new patented family of HIGHSULF™ amine-based desulphurization technologies and strategies was introduced in the literature. Among other applications, the HIGHSULF technology enhances the well-known SCOT or similar processes, uses identical equipment, requires almost no additional capital investment, and produces extremely high quality Claus feed with generic MDEA, or MDEA spiked with promoters. The beauty of HIGHSULF is its simplicity—any existing AGE or SCOT unit can be switched easily to HIGHSULF and back to conventional treating without shutdown. This paper uses the ProTreat™ process simulator to substantiate the technology and to discuss several cases of both the original and extended HIGHSULF process strategies. Attention focuses on using HIGHSULF with MDEA solvents to upgrade low-quality SRU feed and to produce high H_2S gas from the SRU's TGTU amine section. FLEXSORB or other special solvent performance can be had using HIGHSULF with generic MDEA.

(30f) Bibliography of Dr. Ralph Weiland

Mike Resetarits, Simon Chambers and T.J. Cai, Fractionation Research, Inc., Stillwater, OK

Dr. Ralph Weiland's contributions to chemical engineering have been prodigious. Throughout his entire career he has been at the frontier of research in the fields of distillation, absorption and modeling. His company's ProTreat computer program is now globally well accepted for the simulation of acid gas absorption systems. This presentation will highlight certain articles from Dr. Weiland's career.

(55a) Crude Vacuum Column Wash Section - A New Design Approach

Carlos Trompiz, Fluor, Houston, TX

Process simulations using the results of a model developed by Trompiz and Fair at the Separations Research Program (SRP - University of Texas at Austin) for predicting liquid entrainment from sprays can closely reproduce the actual operation of the wash section of crude vacuum distillation columns. The process simulation should use the net wash oil flow rate which reaches the wash bed and not the total wash oil flow rate measured and controlled in the crude vacuum unit operation. This new approach of simulating a crude vacuum column provides the right basis for new designs or revamps involving the wash section of crude vacuum units to prevent coke formation in the wash bed and ensure its proper performance.

(55b) The Impact of Joint Rotation Angle on the Performance of Mellapak N250Y

Michael Basdenand R. Bruce Eldridge, Chemical Engineering, The University of Texas at Austin, Austin, TX

Structured packing elements are typically installed such that the channels of an element are perpendicular to the element above. Previous studies by Owens*et al.*have shown that this leads to significant recirculation and stagnation zones for single phase flows; increased pressure drop was also observed.[1] In this study, different arrangements of Mellapak N250Y were tested using computational fluid dynamics (CFD) to predict the dry pressure drop as a function of the joint rotation and gas flowrate. The computational predictions for the packing elements oriented perpendicular to one another were validated by experimental data.

1. Owens, S. A.; Kossmann, A.; Farone, J.; Eldridge, R. B.; Ketcham, R. A., Flow Field Visualization in Structured Packing Using Real Time X-ray Radiography. *Industrial & Engineering Chemistry Research***2009,**48, (7), 3606-3618.

(55c) Testing of MellapakPlus 352.Y At Fractionation Research, Inc

Marc B. Wehrli, Sulzer Chemtech AG, Winterthur, Switzerland and **Mark W. Pilling**, Sulzer Chemtech USA, Tulsa, OK

Sulzer MellapakPlus^{TM}352.Y packing was tested at Fractionation Research, Inc. in 2011. This high performance packing showed excellent results in the FRI test systems. Test results and interesting comparisons with other high performance packings will be discussed and analyzed.

(55d) FRI Test Data and Performance Modeling of Montzpak B1-250MN

Bjoern Kaibel[1], Helmut Jansen[1], Thomas Rietfort[1], Egon Zich[1] and Zarko Olujic[2], (1)Julius Montz GmbH, Hilden, Germany, (2)Process & Energy Laboratory, Delft University of Technology, Delft, Netherlands

Results are presented of the total reflux distillation experiments carried out with Montzpak B1-250MN at FRI, in a 1.22 m ID column using o/p xylene at 0.1 bar and cyclohexane/n-heptane (CH/n-H) system at 0.31 and 1.62 bar, respectively. All experiments were performed using a Montz high turndown liquid distributor, which performed accordingly over a very wide range of specific liquid loads.

As observed, B1-250MN, the flagship of the newest generation of Montz high performance corrugated sheet structured packings, utilizing perforated sheets with characteristic shallow embossed texture in combination with a reduced corrugation inclination angle and a proprietary smooth bend on lower end of corrugations, exhibits a stable and highly efficient performance over the whole range of operating conditions as encountered in FRI total reflux distillation tests. Interesting to mention is that efficiency curve measured with close boiling, o/p-xylene system matches perfectly that measured under same operating pressure at Bayer TS (BTS) using chlorobenzene/ethylbenzene in a 0.59 m ID column. In both cases the same maximum useful capacity was reached, while the pressure drop observed in FRI test was significantly lower than that measured at BTS. With CH/n-H system slightly lower efficiencies were measured, compared to close boiling system. Practically same efficiencies were obtained at vacuum and above atmospheric pressure, with the pressure drop and maximum useful capacity distributed accordingly. In all cases a strikingly good performance (HETP = 0.25 m at pressure drop below 1 mbar/m) was observed at lowest column loads, and a sudden departure in efficiency was observed upon reaching the point of onset of flooding.

Delft model, with full utilisation of the original correlation tor turbulent vapour flow mass transfer coefficient and a physically founded enhancement of the flow direction change related pressure drop proved to be capable of approaching closely measured efficiency and pressure drop both qualitatively and quantitatively.

Paper No. 55e, Performance Comparison of Structured Packing HETP Correlations

A.K. Jammula and J.R. Whiteley
School of Chemical Engineering
Oklahoma State University

T.J. Cai and M.R. Resetarits
Fractionation Research, Inc.
Stillwater, OK

KEYWORDS

distillation structured packing HETP literature correlations

ABSTRACT

This paper presents an analysis and comparison of five popular HETP correlations for structured packing. The performance comparisons were generated using a screened database containing all of the published HETP experimental results for traditional (not high-capacity) structured packing. The database includes FRI test results which have been released to the public. The screening process excluded data points that fall outside the preloading region. The full database consists of 809 points obtained from 8 different sources. After screening, a total of 766 points were retained for performance evaluation of the different HETP correlations. The total number of packings included in the database is 24. Both hydrocarbon and aqueous systems are represented in the database.

HETP prediction performance is documented for the following correlations: Harrison and France - 1989 [1], Billet - 1999 [4], SRP - 1996 [5], Hanley - 2011 [6] and Delft - 2011 [7]. Performance comparisons are provided using several different correlating factors, e.g. packing type, system type, etc. The paper concludes with a brief discussion of the overall findings.

INTRODUCTION

Accurate HETP predictions are required for proper design and operation of packed columns. Numerous researchers have published HETP research data and results for a variety of structured packing types and operating systems. We have collected all of the published data for traditional structured packing tests (high-capacity structured packing excluded). The full database consists of 809 points obtained from 8 different sources. The complete database was screened to remove 43 data points that fall outside the preloading region. The remaining 766 data points were retained to evaluate performance of five HETP prediction correlations. The distribution of retained data points in the correlation evaluation database is shown in Table 1.

Table 1. Traditional Structured Packing HETP Measurements Published in the Open Literature that Lie Within the Preloading Region

Literature Data Sources	Data Points
SRP	302
Prof. Billet	163
TU Delft	34
Koch	39
Norton	90
FRI	102
Sulzer	11
Spiegel	25
Total no. of points	**766**

The retained data cover 24 different packings in columns ranging from 0.1 – 1.2 m in diameter. Column pressure ranged from 0.021 – 4.14 bar. The following binary systems are represented in the database:

C6 (cyclohexane) / C7 (iso-heptane)
n-octane / toluene
CB (chlorobenzene) / EB (ethylbenzene)
o-xylene / p-xylene
methanol / ethanol
EB (ethylbenzene) / styrene
polyglycol / water

Aspen Plus was used to generate physical properties in situations where physical properties were not reported.

CORRELATIONS

One shortcut and four fundamental HETP prediction correlations were evaluated. Each of the correlations is described briefly below.

Shortcut Correlation

Harrison and France - 1989 [1]

For troubleshooting purposes, Harrison proposed a rule of thumb for both random and structured packings. According to Harrison, structured packing HETP is fixed for a fixed crimp height. According to his rule, HETP is 9 inches for ¼ - inch crimp height, 18 inches for ½-inch crimp and 33 inches for 1-inch crimp. This rule in equation form is

$$HETP = \frac{100}{a_p}$$

This rule is only applicable to packings with a 45° corrugation angle.

Fundamental Correlations

SRP – 1996 [5]

The SRP correlation was developed based on the two-resistance approach. The prediction of HETP requires k_g, k_L and a. In the SRP model, the k_g correlation was developed based on a wetted wall column analogy, the k_L correlation was developed based on penetration theory, and the a correlation was developed based on the earlier work of Shi & Mersmann [8]. SRP model requires a packing specific constant F_{SE} and crimp dimensions for the estimation of effective mass transfer area.

$$k_G = 0.054 \frac{D_G}{S} \left[\frac{\rho_G S\,(u_{Le} + u_{Ge})}{\mu_G}\right]^{0.8} \left(\frac{\mu_g}{D_g \rho_g}\right)^{0.33}$$

$$k_L = 2\sqrt{\frac{0.9 D_L u_{Le}}{\pi S}}$$

$$\frac{a_e}{a_p} = F_{SE} \frac{29.12\,(We_L Fr_L)^{0.15} S^{0.359}}{Re_L^{0.2} \epsilon^{0.6} (1 - 0.93\cos\gamma)(\sin\theta)^{0.3}}$$

$$HETP = \frac{\ln\lambda}{\lambda - 1}\left(\frac{u_G}{a_e\,k_G} + \lambda\frac{u_L}{a_e\,k_L}\right)$$

Billet - 1999 [4]

Billet proposed an HETP correlation which can be applied up to the flooding region. The correlation was developed based on the penetration theory and requires packing specific constants. The correlation can be applied to both random and structured packings. Packing specific constants for various random and structured packings were reported by Billet.

$$k_G = C_V \frac{1}{(\varepsilon - h_L)^{\frac{1}{2}}} \frac{a^{\frac{1}{2}}}{d_h^{\frac{1}{2}}} D_V \left(\frac{u_V}{a\,\nu_V}\right)^{\frac{3}{4}} \left(\frac{\nu_V}{D_V}\right)^{\frac{1}{3}}$$

$$k_L = c_L 12^{\frac{1}{6}} \sqrt{\frac{u_L D_L}{h_L d_h}}$$

$$\frac{a_{Ph}}{a} = 1.5\,(a\,d_h)^{-0.5} \left(\frac{u_L d_h}{\nu_L}\right)^{-0.2} \left(\frac{u_L^2\,\rho_L\,d_h}{\sigma_L}\right)^{0.75} \left(\frac{u_L^2}{g d_h}\right)^{-0.45}$$

$$HETP = \frac{\ln\lambda}{\lambda - 1}\left(\frac{u_G}{a_{ph}\,k_G} + \lambda\frac{u_L}{a_{ph}\,k_L}\right)$$

Hanley - 2011 [6]

Hanley developed an HETP correlation by utilizing the point HETPs in the column, unlike other correlations where an average value was used for the calculation of HETP. Hanley did not make any assumptions for the development of the k_g, k_L and a correlations. These correlations were developed using the Buckingham pi theorem and the Chilton & Colburn analogy [9]. The correlation does not require packing specific constants. This correlation is limited to the preloading region.

$$k_L = 0.33\, Re_L Sc_L^{\frac{1}{3}} \left(\frac{c_L D_L}{d_e}\right)$$

$$k_G = 0.0084\, Re_V Sc_V^{\frac{1}{3}} \left(\frac{c_V D_V}{d_e}\right)\left(\frac{\cos\theta}{\cos\frac{\pi}{4}}\right)^{-7.15}$$

$$\frac{a_m}{a_d} = 0.539\, Re_V^{0.145} Re_L^{-0.153} We_L^{0.2} Fr_L^{-0.2} \left(\frac{\rho_V}{\rho_L}\right)^{-0.033} \left(\frac{\mu_V}{\mu_L}\right)^{0.090} \left(\frac{\cos\theta}{\cos\frac{\pi}{4}}\right)^{4.078}$$

$$HETP = \frac{G}{a_m}\left(\frac{C_y}{k_y} + \frac{C_x}{k_x}\right)$$

Delft – 2011 [7]

The Delft correlation was first developed in 1999 and has been routinely updated as additional Montz experimental data became available. The Delft vapor phase mass transfer coefficient correlation was developed based on transitional flow and the liquid phase mass transfer coefficient correlation was developed from the penetration theory. The Delft effective mass transfer area correlation is a modified Onda correlation [10]. The Delft correlation does not require any packing specific constants, but it does require packing crimp dimensions for the calculation of HETP.

$$k_G = \sqrt{k_{G,lam}^2 + k_{G,turb}^2}$$

$$k_{G,lam} = \left(\frac{D_G}{d_{hg}}\right) 0.664\, Sc_G^{\frac{1}{3}} \sqrt{Re_{Grv} \frac{d_{hG}}{l_{G,pe}}}$$

$$k_{G,turb} = \left(\frac{D_G}{d_{hG}}\right) \frac{Re_{Grv}\, Sc_G \frac{\xi_{GL}\varphi}{8}}{1 + 12.7\sqrt{\frac{\xi_{GL}\varphi}{8}}\left(Sc_G^{\frac{2}{3}} - 1\right)} \left[1 + \left(\frac{d_{hg}}{l_{G,pe}}\right)^{\frac{2}{3}}\right]$$

$$k_L = 2\sqrt{\frac{D_L u_{Ls}}{\pi\ \varepsilon\, h_L 2s}}$$

$$\frac{a_e}{a_p} = (1 - \Omega)\left\{1 - \exp\left[-1.45\left(\frac{0.075}{\sigma}\right)^{0.75} Re_{Ls}^{0.1} Fr_{Ls}^{-0.05} We_{Ls}^{0.2}\right]\right\}$$

$$HETP = \frac{\ln \lambda}{\lambda - 1} \left(\frac{u_G}{a_e\, k_G} + \lambda \frac{u_L}{a_e\, k_L} \right)$$

RESULTS

Predictive performance, reported as mean absolute relative error (MARE), of each of the five correlations using the SRP experimental data are summarized in Table 2.

Table 2. Summary of Literature Correlation Performance for SRP HETP data

Packing	Harrison Deviation %	SRP Deviation %	Billet Deviation %	Hanley Deviation %	Delft Deviation %
Flexipak 2	8	23	10	26	24
Gempak 2A	21	30	10	28	22
Intalox 2T	33	29	15	14	10
Maxpak	43	22	18	11	8
Montz B1-250.60	13	19	11	45	15
Montz B1-250	24	24	26	17	12
Montz B1-350	15	21	36	13	20
Montz B1-400.60	26	32	20	37	17
Montz B1-400	8	29	34	10	33
Montz BSH-400.60	25	36	15	39	17
Montz BSH-400	7	20	20	10	29

Literature correlation performance for all databases with parity plots will be presented in the AIChE meeting. The full paper will include a discussion of the overall results.

Bibliography

1. Harrison, M., Distillation Column Troubleshooting, Part 2: Packed Columns. Chem. Eng. 1989, 121.
2. Kister, H. Z., Distillation Design. McGraw-Hill: 1992.
3. Lockett, M. J., Easily predict structured-packing HETP. Chem. Eng. Prog., 1998, 94(1), 60–66.
4. Billet, R.; Schultes, M., Prediction of Mass Transfer Columns with Dumped and Arranged Packings: Updated Summary of the Calculation Method of Billet and Schultes. Chemical Engineering Research and Design 1999, 77 (6), 498-504.
5. Rocha, J. A.; Bravo, J. L.; Fair, J. R., Distillation Columns Containing Structured Packings: A Comprehensive Model for Their Performance. 2. Mass-Transfer Model. Industrial & Engineering Chemistry Research 1996, 35 (5), 1660-1667.
6. Hanley, B.; Chen, C.-C., New mass-transfer correlations for packed towers. AIChE Journal 2011.

7. Olujic, Z. In Predicting the Liquid-Side Mass-Transfer Resistance of Structured Packings: State of the Art, AIChE, Chicago, Illinois, 2011; p 71b.
8. SHI , M., Mersmann A., Effective interfacial area in packed columns. German Chemical Engineering 1985, vol. 8 (no2), pp. 87-96.
9. Chilton, T. H.; Colburn, A. P., Mass Transfer (Absorption) Coefficients Prediction from Data on Heat Transfer and Fluid Friction. Industrial & Engineering Chemistry 1934, 26 (11), 1183-1187.
10. Onda, K., Mass Transfer Coefficients Between Gas And Liquid Phases In Packed Columns. Journal of Chemical Engineering of Japan 1968, 1 (1), 56-62.

(55f) Packing Structure Capable of Sub-Centimeter HETP for High Purity Distillation of Low Relative Volatility Compounds

Ward E. Tegrotenhuis, Pacific Northwest National Laboratory, Richland, WA and Michael R. Powell, Thermal & Reaction Systems, Pacific Northwest National Laboratory, Richland, WA, US Virgin Islands

Achieving a high separation factor with mixtures of low relative volatility typically requires a large numbers of stages and very tall columns, even with advanced structured packings. The challenge is exacerbated by vapor and liquid phase dispersion when a high purity product is demanded. A novel microchannel technology is presented that utilizes layered structures to precisely control the mass transfer length-scale in the liquid phase to a fraction of a millimeter, thereby enabling sub-centimeter height of a theoretical plate (HETP). Furthermore, both the liquid and vapor phase flows are laminar, which reduces dispersion and allows high purity to be achieved even with low relative volatility mixtures. Experiments conducted with low concentrations of 1,4-dimethyl cyclohexane and 3-methyl heptane in octane successfully demonstrated HETP values below 0.5 cm for relative volatilities of 1.03 and 1.24, respectively. The lack of HETP dependence on relative volatility is evidence that dispersion is not limiting product purity. Scale-up of the technology is discussed.

COMBINATION FILTRATION OF PRESSURE, VACUUM AND CLARIFICATION TECHNOLOGIES FOR OPTIMUM PROCESS SOLUTIONS

Barry A. Perlmutter, President & Managing Director
BHS-Filtration Inc.
9123-115 Monroe Road
Charlotte, North Carolina 28270
Phone: 704.845.1190
Fax: 704.845.1902
E-mail: barry.perlmutter@bhs-filtration.com

INTRODUCTION

Filtration experts, over the years, have discussed combination filtration and debated its definition. In the realm of cartridge filtration, simply defined, a combination filter is one that does at least one other processing job at the same time as filtering a suspension. For example, this could be carbon canister which removes both suspended and dissolved components. In water applications, a combination filter removes bacteria, sediment, chlorine taste and odor, and scale. In lubrication oil filtration, combination filtration refers to full-flow and by-pass flow filtration. Finally, for small scale process filtration, combination filtration is installing bag and cartridge filtration systems in series.

There is, however, a new definition of combination filtration that transcends the standard approach and will assist process engineers with trouble shooting and "idea-generation." The approach relies upon the slurry analysis and testing to uncover the "process symptom" and then develop a process solution called "combination mechanical slurry conditioning and filtration." This article discusses slurry testing and process analysis and then illustrates seven (7) case history example installations showing "combination mechanical slurry conditioning and filtration."

TESTING TO DETERMINE "COMBINATION MECHANICAL SLURRY CONDITIONING AND FILTRATION"

Overview of Bench Top Testing in the BHS Laboratory

The BHS bench top testing is conducted using the BHS Pocket Leaf Filter, as shown in Figure 1. The testing will analyze cake depths, operating pressures, filter media, washing and drying efficiencies and qualitative cake discharge. The data collection sheets are shown in Figure 2.

Typical Example of "Combination Mechanical Slurry Conditioning and Filtration" from Bench Top Testing in the BHS Laboratory

The typical testing program is shown in Figure 3. The results and summary of the testing illustrated that the most effective approach for the process filtration is a two stage process.

The primary filtration is by candle filters on a continuous basis. These filters are able to retain particles down to 0.5 microns and discharge these solids as a concentrated sludge/slurry. In order to reduce utilities, no drying or washing is necessary in the larger candle filters.

The concentrated slurry will then be filtered on a secondary pressure plate filter to recover the remaining solvent and discharge the dry solids. The secondary filtration is a batch process which will incorporate filtration, counter current washing, drying, and dry cake discharge. The overall result is a very dependable process with high quality filtrate and minimal utility air and water usage. Each of these processes for the primary and secondary filtration has been tested to confirm that the overall design requirements can be met.

COMBINATION FILTRATION TECHNOLOGIES OF PRESSURE, VACUUM AND CLARIFICATION

There are, without doubt, many technologies that can be applied in combination, including the use of chemicals such as flocculants and coagulants, and which would be the subject of a filtration-chemical handbook/textbook. These already exist in the marketplace. However, from a practical viewpoint, for the basis of this article, a brief description of five (5) types of filtration technologies is presented below based upon general operating conditions at chemical plants.

High-Solids Slurries: Continuous Pressure Operation

The Rotary Pressure Filter technology provides for thin-cake, continuous production in a single unit. Filtration is conducted via pressure of up to 90 psig. Positive displacement washing or counter-current washing follows filtration. Of course, multiple washing steps as well as solvent exchanges, steaming and extraction can also be accomplished. Finally, the cake is dried by blowing hot or ambient-temperature gas through the cake. The Filter has a uniquely designed discharge system, which provides for atmospheric discharge from pressure filtration. After automatic cake discharge, the filter cloth is washed; the clean filter cloth the re-enters the feeding / filtration zone thereby continuing the process. All solvent and gas streams can be recovered separately and reused in the process to minimize their consumption.

As for the operation, each process zone (typically 5 - 7 zones) is isolated by a separating element. The pressure in the specific zone can be adjusted to meet the process parameters. The separating elements are sealed to the rotating drum via a gas membrane pressure seal. This drum is sealed to the outer housing by drum packing. The drive system with self-sealed roller bearings provides for high stability and low stress on the drive.

High-Solids Slurries: Continuous Vacuum Operation

High-solids slurries can be defined as up to 50 – 55% solids in the slurry feed. In some cases, for high solids applications, the slurries can be better handled using vacuum filtration rather than pressure filtration. An example of a continuous, thin-cake technology is a Continuous – Indexing Vacuum Belt Filter. This technology consists of fixed vacuum trays, continuously feeding slurry system and indexing or step-wise movement of the filter media. The filter media is indexed by pneumatic cylinders located on the exterior of the unit. The pneumatic operation and fixed trays eliminates a motor and variable speed drive, there are no rails/rollers, and no rubber carrier belt.

Low-Solids Slurries for Clarification & Recovery: Batch Pressure Operation

Candle Filters and Pressure Plate Filters are installed for clarification and recovery applications from liquids with low solids content. These units offer full containment, fines removal in the 0.5 micron and finer range, and can be fully automated. The candle filters are vertical candles while the pressure plate filters are horizontal plates. The major difference between the two units depends on the cake structure that is formed. Some cakes are better handled in the horizontal and some in the vertical. Cake discharge is automatic either by gentle gas expansion of the filter sock in a candle filter or via plate vibration with gas-assist for the pressure plate filter.

Low-Solids Slurries for Clarification & Recovery:
Conventional versus Contained Filter Presses

Conventional filter presses are open design with a series of vertically mounted plates. The Filter cloth is hung over each plate, extending from top to bottom, and is held in place by eyelets that fit over the cloth-pins on top of the plates. The plates are compressed by hydraulic closing pressure. Cake discharge can be manual or automatic.

Contained filter presses consist of circular filter plates that are sealed in a movable housing. The housing is sealed to 150 psig and full vacuum. The housing is then contained in an outer enclosure. The filter plate design allows for forward and reverse flow operations and can be automatically cleaned to less than 1 ppm residual cross contamination. Process steps can be pressure or vacuum and cake discharge is automatic using scraper knives.

CASE HISTORY EXAMPLES SHOWING "COMBINATION MECHANICAL SLURRY CONDITIONING AND FILTRATION"

The next sections review various actual installations showing the benefits of "combination mechanical slurry conditioning and filtration." These examples are included to illustrate creative idea-generation when examining a process problem.

Continuous Vacuum Filtration Followed by Candle Filtration

In this case, shown in Figure 4, the slurry contained a high solids loading but with a very small particle size distribution. Continuous vacuum filtration solved the first part of the process. The cake is very sticky with a cake thickness of 5 – 6 mm. The mother filtrate contained small, very fine solids. These solids were filtered, washed, and concentrated in candle filters and then sent back to the process.

Secondly, as the solids are sticky, the cloth wash contains solids. These solids, as above, are filtered and concentrated also in candle filters and then sent back to the process.

Candle Filtration Followed by Continuous Vacuum Filtration

In this case, shown in Figure 5, the candle filters operate to mechanically condition the slurry through thickening and concentrating up to 10% solids. The resulting slurry can then be economically and technically processed on a vacuum belt filter by vacuum filtration, cake washing and drying. The drying can be by vacuum, compression, blowing with hot or ambient-temperature gas or steaming. Applications for this approach occur in chemical plants as well as in coal gasification plants.

Candle Filtration Followed by Pressure Plate Filtration

This is the case described in the testing section "Typical Example of "Combination Mechanical Slurry Conditioning and Filtration" and shown in Figure 3. The process flow is shown in Figure 6.

Continuous Vacuum Filtration Followed by Contained Filter Press Filtration

In this process, there are three process reaction steps occurring which are being processed by only two types of filtration technologies. Figure 7 illustrates the process flow.

The first reaction requires vacuum filtration, cake washing and drying by vacuum only; the crystals are fragile and cannot be mechanically pressed. The continuous-indexing vacuum belt filter can have the pressing device installed but not operated.

The second step has more robust crystals and requires a lower moisture specification. The cake from the second reaction and is then processed on the same vacuum belt filter but, in this case, the cake is mechanically compressed with blowing for final drying.

The cake from the second step undergoes further processing, but, in this case, pressure filtration is the optimum solution. Due to the nature of the chemicals and solvents, a contained filter press for pressure filtration, cake washing and drying, by both blowing and vacuum is accomplished.

Continuous Pressure Filtration Followed by Candle Filtration

For this commodity chemical, with rates up to 60 tons of dry solids/hour, the initial mother liquor contains fines in the 1 – 5 micron size range. The process decision is to accept fines in the mother liquor rather than have a reduced rate with a tighter filter cloth. The fines are then recovered in the candle filters which can either produce a dry cake or concentrated slurry. The flow diagram is shown in Figure 8.

Continuous Vacuum Filtration (Two Variations)

In this case, similar to one of the previous examples, it was more efficient to use the same vacuum belt filter with two different variations and to conduct the initial cake washing in a separate agitated reslurry tank. Figure 9 shows this process.

Candle Filtration Followed by Conventional Filter Press

In some applications, a conventional filter press is the correct technology but the initial slurry feed is the problem. For these applications, mechanically conditioning of the slurry by concentrating candle filters is the optimum approach. The concentrating candle filters will increase the solids concentration in the slurry as well as reduce the volume of slurry that is to be treated by the filter press. Specialty chemical, chlor-alkali and coal gasification plants are typical examples. Figure 10 shows this flow process.

SUMMARY

This article discussed a new definition of combination filtration that will provide process engineers a framework for "idea-generation" when analyzing an operating bottleneck. The approach relies upon the slurry analysis and testing to uncover the "process symptom" and then develop a process solution called "combination mechanical slurry conditioning and filtration." This article illustrated that the installation and combined use of filtration technologies, while higher in capital cost, will result in a more reliable operating process at the plant. In all cases, however, only through accurate lab and pilot testing, will the optimum filter selection be realized.

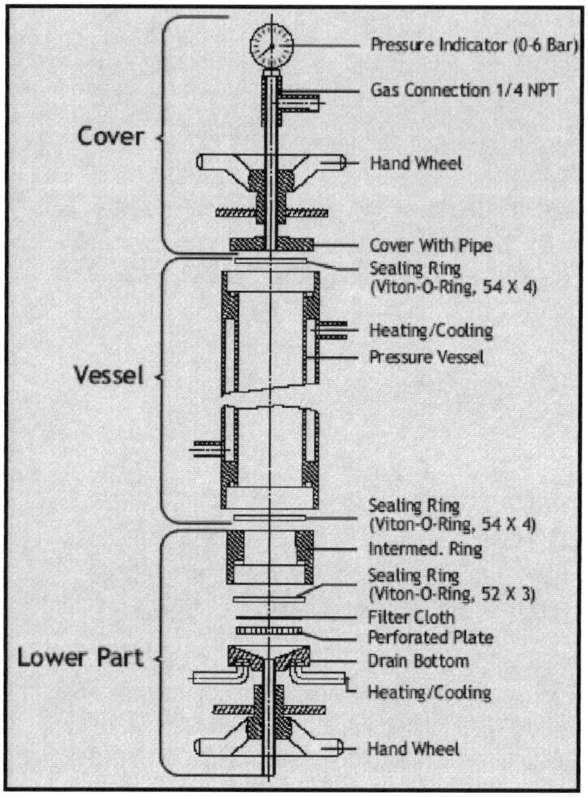

Figure 1: BHS Pocket Leaf Filter (PLF)

Customer :		Test Number:
Date :		Run #
	Filter Media	
	Suspension	
Filling	Volume of Slurry	
	Density of Slurry	
	% Solids in Feed	
	Temperature	
Filtration	Vacuum or Pressure	
	Volume of Filtrate	
	Time for Filtration	
	% Solids in Filtrate	
Wash 1	Wash Material	
	Pressure	
	Volume of Filtrate	
	Time for Filtration	
Wash 2	Wash Material	
	Pressure	
	Volume of Filtrate	
	Time for Filtration	
Drying	Pressure	
	Temperature	
	Flow Rate	
	Time for Drying	
	Pressing Pressure	
Cake	Weight	
	Thickness	
	% Residual Moisture	
	Dry Cake Weight	
	Cake Discharge OK?	

Figure 2: Data Collection Sheet for BHS Pocket Leaf Filter (PLF)

1. Test Purposes

The purposes of these tests were to:

1.1 Determine required filter area for the primary filtration.
1.2 Evaluate filtrate quality of both primary and secondary filtration.
1.3 Determine filter aid requirement for secondary filtration.
1.4 Evaluate washing efficiency and minimize amine loss in filter cake.

2. Test Methods

The BHS-Filtration Pocket Leaf Filter, jacketed with a 400 ml capacity and with 20 cm^2 filter area, was used to gather data and make observations on this product.

The following information was gathered during this test:

Primary Filtration Test Methods:

2.1 Variable speed peristaltic pump was used to conduct primary filtration tests.
2.2 Filtration Rate vs. Differential Pressure
2.3 Filtrate Quality vs. Filter Media

Secondary Filtration Test Methods:

2.4 Filtration pressure for secondary filtration tests was provided by compressed air.
2.5 Filtration Rate vs. Filter-Aid Concentration
2.6 Filtrate Quality vs. Filter-Aid Concentration
2.7 Filtration Rate vs. Filter Media
2.8 Filtrate Quality vs. Filter Media
2.9 Filter Cake Residual Amine (Filtrate Conductivity) vs. Volume Wash Water
2.10 Filter Cake Residual Moisture vs. Drying Air Usage

Figure 3: Typical Testing Program

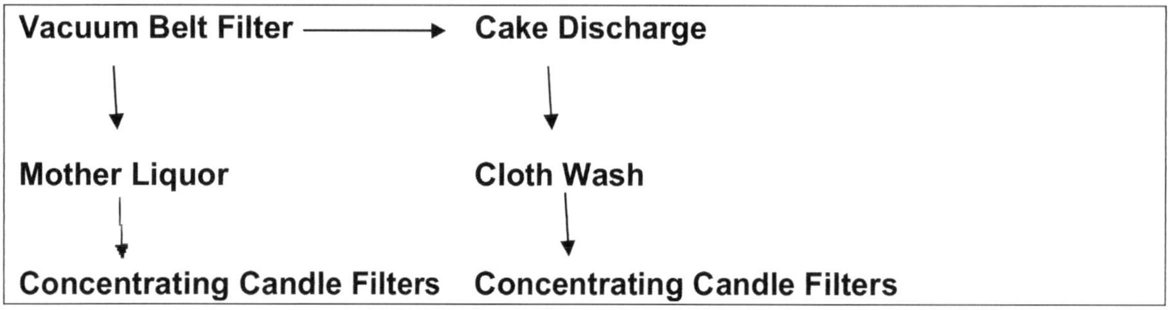

Figure 4: Continuous Vacuum Filtration Followed by Candle Filtration

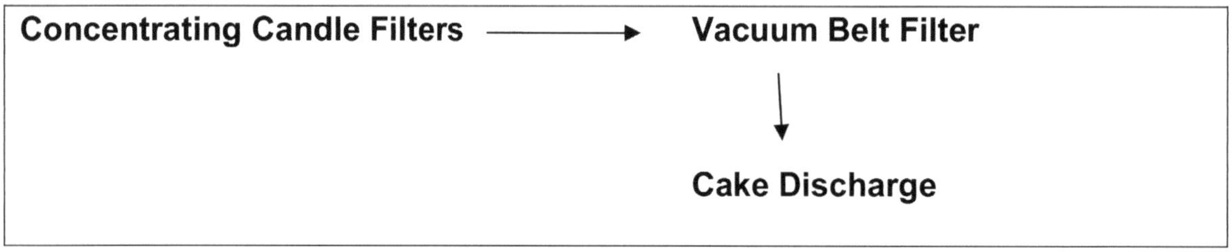

Figure 5: Concentrating Candle Filters Followed by Continuous Vacuum Filtration

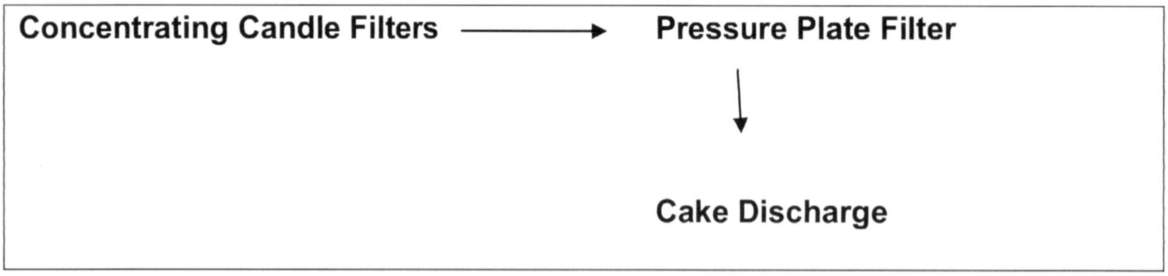

Figure 6: Concentrating Candle Filters Followed by Pressure Plate Batch Filtration

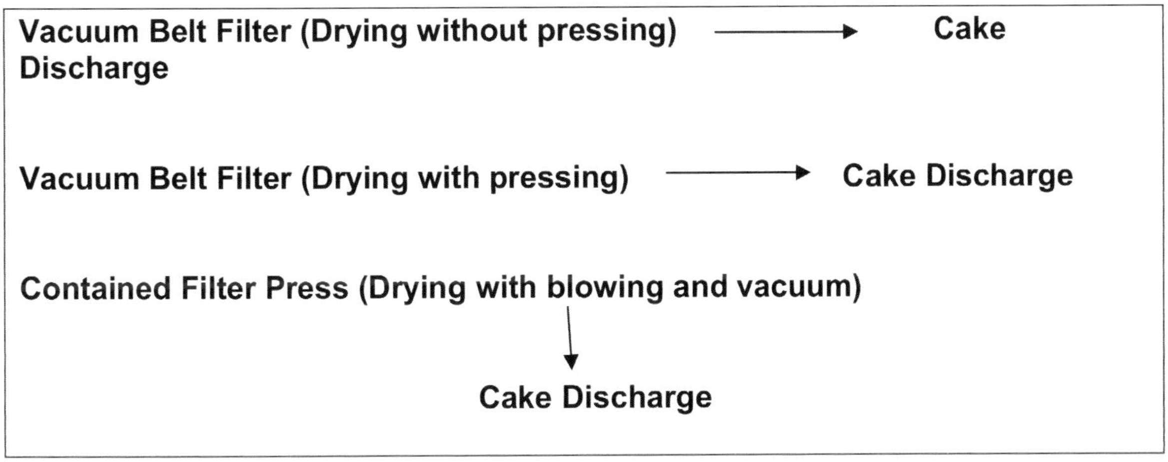

Figure 7: Continuous Vacuum Filtration (Two Variations) Followed by Contained filter Press

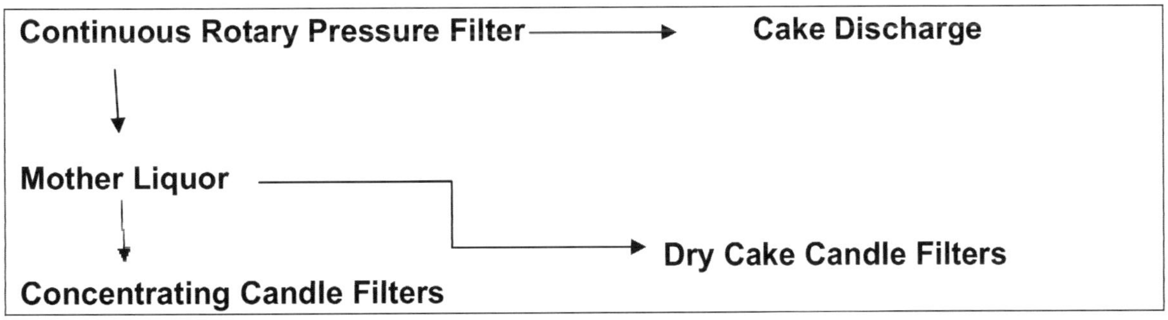

Figure 8: Continuous Rotary Pressure Filtration Followed by Candle Filtration

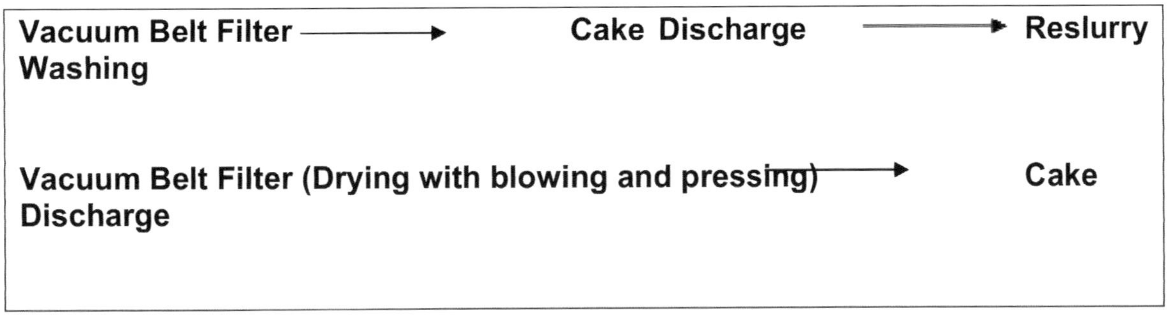

Figure 9: Continuous Vacuum Filtration (Two Variations)

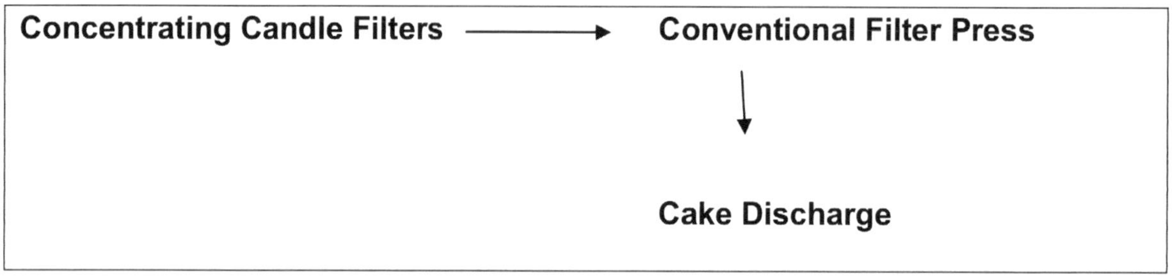

Figure 10: **Concentrating Candle Filters Followed by Conventional Filter Press Batch Filtration**

(53b) Prediction of Salt Effect on Vapor - Liquid Equilibrium of Methyl Acetate - Methanol System

Dhanalakshmi J., Sai P.S.T. and Balakrishnan A.R., Chemical engg, IIT Madras, chennai, India

The salting coefficient of salt in polar system (methyl acetate - methanol) is obtained by using scaled particle theory has been used. The scaled particle theory is a purely predictive method for evaluation of the salt effect on vapor - liquid equilibrium of methyl acetate - methanol at 1 atm. It requires only molecular parameters for calculating the salt effect. From scaled particle theory, the salt effect of polar molecule have examined in non-aqueous system. In the present study, the salting coefficient of polar molecule is predicted for various inorganic salts. The effect of cation and anion on the salting out phenomenon of methyl acetate - methanol system is reported. It is observed that salting coefficient is depending on the size and polarizability of the ions in non-aqueous solution. Therefore, the salting coefficient for a ternary system of methyl acetate - methanol - salt is well determined by scaled particle theory.

Keywords: methyl acetate, methanol, scaled particle theory, salting coefficient

(53d) Design of a Thermal Separation Process for Draw Solution Recovery in Forward Osmosis Desalination Plant

Tae-woo Kim, Department of Chemical and Biomolecular Engineering, Korea Advanced Institute of Science and Technology, Daejeon, South Korea, Young Kim, Energy Plant Research Division, Korea Institute of Machinery and Materials, Daejeon, South Korea, Woohyun Kim, Korea Advanced Institute of Science and Technology, Department of Chemical and Biomolecular Engineering , Daejeon, South Korea and Sunwon Park, Korea Advanced Institute of Science and Technology, Department of Chemical and Biomolecular Engineering, Daejeon, South Korea

The optimal design of a draw solution recovery process for a forward osmosis desalination plant is studied. The previous researches in forward osmosis (FO) desalination are mainly focused on the water permeability and salt rejection of semi-permeable membrane. However, relevant engineering issues such as recovery of draw solution and energy efficiency of process should be also resolved to commercialize a FO desalination process.

In this study, a continuous process is designed using a distillation column to separate draw solution and permeated water. The optimal design of the process is suggested achieving the minimum production cost while purity of product water is maintained at drinkable level. For case studies, ammonium bicarbonate, ammonium hydroxide and ethanol are employed as draw solutes which are selected by the screening method based on property database in our previous research. Each draw solute system is individually designed to handle redissolving or reconcentrating problems of each case.

(53e) Leaching of NdFeB Magent Scarp

Ho-Sung Yoon, Korea Institute of Geoscience and Mineral Resources, Daejeon, South Korea and Kyeong Woo Chung, Metal Recovery Department, Korea Institute of Geoscience and Mineral Resources, Daejeon, South Korea

Leaching of NdFeB magent scarp

Ho-sung Yoon*, Kyeong Woo Chung, Chul-ju Kim

Metal Recovery Dept., Korea Institute of Geoscience and Mineral Resources, Daejun 305-350, Korea

*E-mail: hsyoon@kigam.re.kr, Tel: +82-42-868-3611

Currently, large amounts of scarp form NdFeB magnet manufacture are stockpiled owing to the lack of a cost-effective processing method. Because of the growing new market for individual rare-earth metals such as Nd and owing to the likely continuing increase in demand for permanent magnets, scrap treatment will undoubtedly become an important alternative materials source. Development of a viable method to process NdFeB scarp will aid manufacturers economically and reduce the necessity of consuming rare-earth materials.

In the present study, Nd was recovered from NdFeB scrap. To dissolution Nd and Fe, scrap was oxidizing-roasted and leached in sulfuric acid solution. Nd and Fe was separated using double-salt precipitation method by the addition of sodium sulfate to leach liquor. Optimized temperature of oxidizing-roasting was 500°C for sintering magnet and 700°C for bonded magnet. When oxidizing-roasted NdFeB magnet was treated in 2M sulfuric acid at 50°C for 2hr with the pulp density of 15%, 99.4% of Nd and 95.7% of Fe was leached and recovered.

() Musings on Management

Frank van Lier, The Lubrizol Corporation, Wickliffe, OH

It really is all about the people!

Hiring good people.

Retaining them.

Adjusting for different personal styles, cultures, encouraging diversity, embracing different perspectives.

Motivating people to work together in successful global teams.

Achieving top notch results that are better than any individual can achieve alone.

Maintaining the focus.

Getting the results.

Management is all about the excitement managers gets from having their people and teams grow and succeed.

(94a) Multichordal Gamma Scans with Quantitative Analysis for Diagnosing Tray Maldistribution

Henry Kister, Process Engineering, Fluor Corp., Aliso viejo, CA

MULTICHORDAL GAMMA SCANS WITH QUANTITATIVE ANALYSIS: KEY TO DIAGNOSING TRAY MALDISTRIBUTION

Henry Z. Kister, Fluor Corp., Aliso Viejo, CA

Large-diameter olefins towers like primary fractionators and C3 Splitters occasionally experience liquid and/or vapor maldistribution in either conventional or high-capacity trays. Such maldistribution often leads to tray efficiencies much lower than design, forcing an increase in reflux and reboil in the tower, and limiting tower capacity. When such maldistribution occurs, it may be difficult to diagnose using conventional troubleshooting techniques such as vendor software, delta P measurement, and conventional single-chord qualitative gamma scanning. Judicious multichordal gamma scanning with quantitative analysis is a very useful tool for diagnosing maldistribution on trays. This paper discusses this technique and describes its application in a couple of case studies.

(94b) A Dying Column, Hidden Damage And The Weeping Trap Syndrome

Sean Hennigan, BP Hull Research and Techonlogy Centre, Saltend, United Kingdom

A deterioration in a petrochemicals column performance offered a challenging puzzle. Investigations were carried-out into the potential root causes, through data analysis and trials, all of which suggested a severe stage loss through the system - this was at odds with the column scan which showed much more restricted damage. Opening the column showed extensive manway loss and panel damage, apparently confirming a simple explanation of poor manway installation. However detailed analysis of other causes led us to highlight a previously underestimated vulnerability with this kind of column named "Weeping Trap Syndrome" or WTS. Cases where the steady state load profiles in the column change severely tray to tray are most susceptible to WTS. Whilst most trays are well loaded others are on the edge of weeping even at design.

Following this theory through, the loss of a few trays can have a significant overall stage loss as the low load front shifts down the column. For most towers the numbers of stages is a threshold function of rate - hit the right load to seal the trays and the stage count is fixed. For a WTS column we postulate an odd feedback mechanism where the number of stages depends on both rate and the degree of separation you achieve. This leads to potentially unworkable startups.

(94c) Effect of Tray Spacing on Tray Performance

Simon Chambers, T.J. Cai, A.N. Vennavelli and A. Y. Ogundeji, Fractionation Research, Inc., Stillwater, OK

At the 2007 AIChE Spring Meeting, FRI presented information on the effect of deck open area and hole size on tray performance. The theme of tray design effects on performance is extended to tray spacing. The impact of three tray spacings on capacity, efficiency, pressure drop and rangeability will be described and illustrated. We will also show that the impact of tray spacing on tray capacity is other than TS**0.5. FRI is currently performing additional research work in air-water and hydrocarbon systems regarding closely-spaced trays.

APPLICATION OF ULTRA HIGH CAPACITY SHELL CONSEP* TRAY FOR DE-BOTTLENECKING OF A CRUDE DISTILLATION UNIT

Kaushik Majumder, Shell Projects & Technology, Bangalore, India
Giuseppe Mosca, Sulzer Chemtech Ltd., Winterthur, Switzerland
Kent Mahon, The New Zealand Refining Co. Ltd., Whangarei, New Zealand

Abstract

The main fractionator of the Crude Distillation Unit (CDU-1) in Whangarei refinery of The New Zealand Refining Company Limited (NZRC) was retrofitted with high capacity internals to increase the unit throughput from 8500 ton/day to 13000 ton/day. Shell ultra high capacity ConSep trays were applied in the most capacity constrained HGO pump-around (mid circulating reflux) section of the column as no other first generation high capacity tray was found adequate to debottleneck this section. By the application of ConSep trays, capex savings of the order of USD 5.5 - 6 million was achieved compared to other conventional debottlenecking options. This was the first application of ConSep trays in a CDU main fractionator and the post revamp test run established the realization of the expected performance.

Introduction

NZRC's Whangarei refinery targeted expanding the refining capacity through 'Point Forward Project' [1]. The project involved increasing the throughput of the CDU-1 from 8500 ton/day to 13000 ton/day, thereby increasing distillate component to downstream processing and generating additional long residue to replace imported long residue for loading the vacuum distiller. Figure-1 shows the simplified process flow diagram of CDU-1.

Shell Global Solutions International (SGSi) carried out the feasibility study for the expansion of CDU-1. Several options were studied to debottleneck the main fractionator:

1. Replacement of the existing column internals with high capacity internals including ConSep tray for the most capacity constrained HGO pump-around section.
2. Installation of a new crude pre-fractionator column to separate off light naphtha and reduce the load to the main fractionator. The capex for this option was found USD 6 million higher than for option 1.
3. Installation of a new heavy end column to recover the Heavy Gas Oil (HGO) dropped into the long residue to offload the main fractionator. The capex for this option was found USD 5.5 million higher than for option 1.

On basis of the comparison of the revamp options, NZRC has decided to proceed with the ConSep tray alternative owing to its lowest capex and most favourable economics.

Shell ConSep Tray Technology

The ConSep tray utilizes the principle of de-entrainment by centrifugal forces to remove the gravitational limitation of jet flood. Separation of the entrained liquid before entering the next tray allows very high vapour velocities to be achieved in the column. The ConSep tray combines the features of a contacting deck and a separator deck in a single tray. The basic features of the ConSep tray are illustrated in Figure-2. The functioning of the contacting deck, which in fact is a normal tray, is limited by three hydraulic mechanisms: jet flooding, downcomer choking and downcomer backup. The use of separator deck influences all three mechanisms [2]:

1. The jet flooding limit is extended as the entrained liquid is efficiently separated from the vapour to prevent carry-over of liquid to the tray above.
2. The liquid entering the main downcomer is largely coming from the separator deck, where it is well degassed. As a result, the downcomer liquid handling capacity is substantially increased.
3. To eliminate downcomer back-up limitation, the separator deck is designed with low pressure drop swirl tubes. The contacting deck is also designed with relatively high open area.

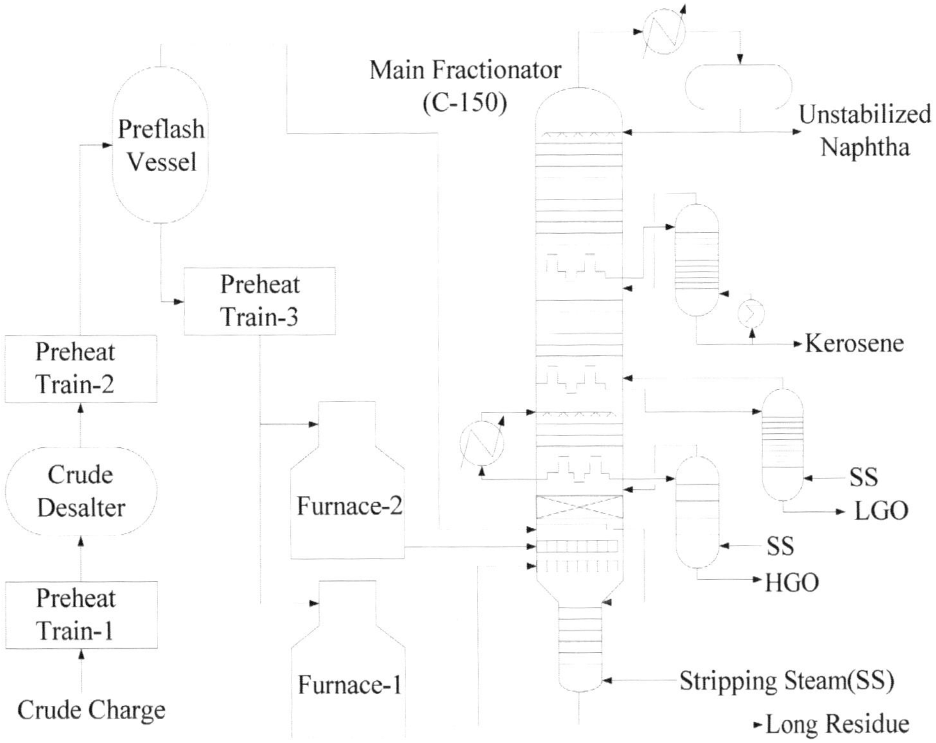

Figure-1: Simplified process flow diagram of CDU-1

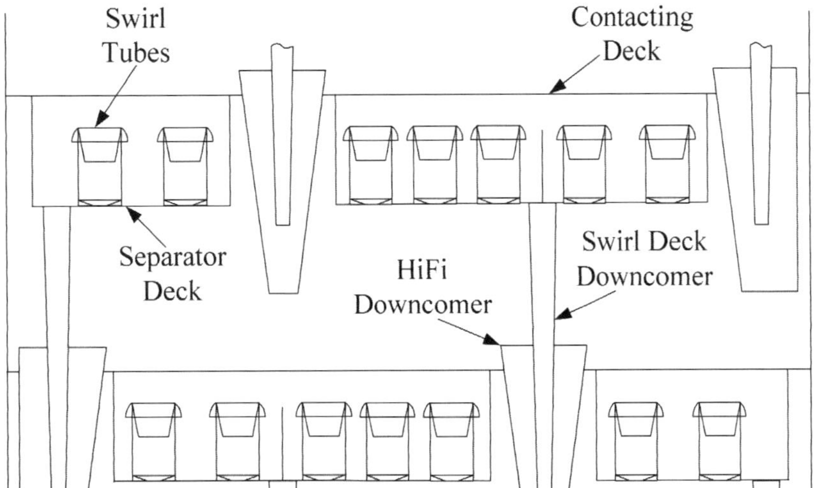

Figure-2: Schematic drawing of ConSep tray

Figure-3 illustrates the expected capacity gain of the ConSep tray over conventional trays and packing [3].

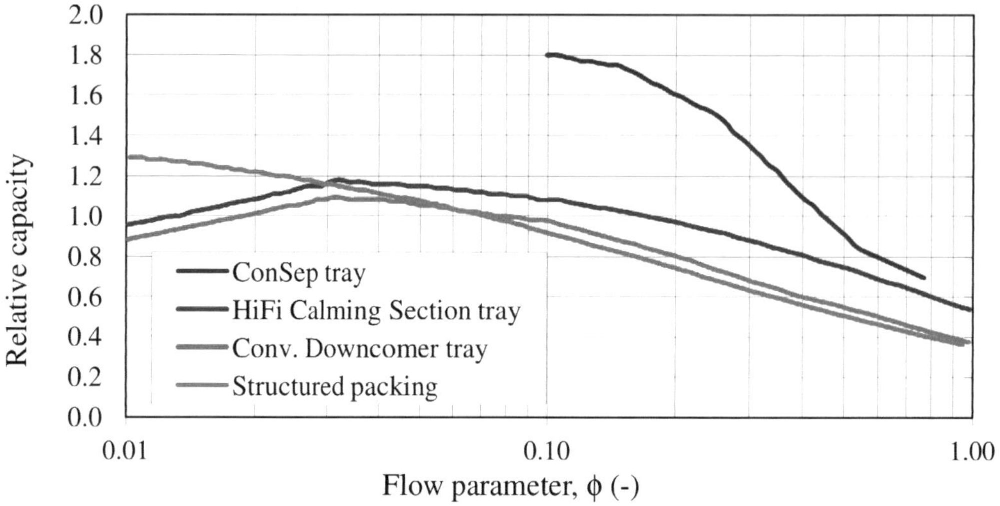

Figure-3: Comparison of column internal capacities

The flow parameter (ϕ) is defined as $\phi = \frac{L}{V}\sqrt{\frac{\rho_V}{\rho_L}}$, where $\frac{L}{V}$ represents the liquid-to-vapour mass flow ratio and $\frac{\rho_V}{\rho_L}$ represents the ratio of vapour-to-liquid density.

Typically the ConSep tray is capable of offering 40-50% capacity advantage over a wide range of first generation high capacity trays. In most revamps, the column retrofitted with the ConSep tray gets limited by other factors such as availability of feed and/or constraints on auxiliary equipment e.g. reboiler, condenser, pumps etc. even before the full potential of the ConSep tray is realized. Table-1 shows some of the applications of the ConSep tray along with benefits achieved and constraints faced.

Table-1: Overview of ConSep tray applications

Location	Diameter	Application	Installed	Capacity Increase, %
UK	1.9	NGL Debutanizer	1995	22 (1)
Australia	1.9	FCCU Debutanizer	1996	30 (2)
Germany	2.2	HCU Main Fractionator	1999	50
Australia	1.7	NGL Debutanizer	1999	15 (1)
Singapore	2.5	FCCU Debutanizer	2000	20 (1)
Japan	2.1	FCC Debutanizer	2006	10
Sweden	2.0	FCC Debutanizer	2006	20
Sweden	1.0	C3/C4 splitter	2007	50
U.S.A.	2.3	C3/C4 splitter	2006	12
Canada	1.0	Depropanizer	2007	20
New Zealand	4.6	Crude Distillation	2009	22 (2)
Singapore	1.8	PO Drying Column	2011	20 (2)
China	3.2	Ethylene Fractionator	2011	Target 25% (3)
Japan	2.5	FCC Debutanizer	2013(scheduled)	Target 15% (3)

(1) limited by reboiler capacity (2) limited by feed to column (3) no data yet

Modifications of Main Fractionator (C-150)

During the feasibility study, the HGO pump-around section of the column was found severely limiting for the targeted throughput of 13000 ton/day. This section was already fitted with Shell CS** trays (Calming Section trays). Since the first generation of high capacity internals were found inadequate to debottleneck this section, ultra high capacity Shell ConSep trays were selected. The trays were designed to achieve 33% more capacity compared to Shell CS tray.

The HGO pump-around section consisted of 3 contacting trays with a tray spacing of 500 mm. A one-for-one tray replacement with ConSep trays was selected. Figure-4 shows a schematic drawing of HGO pump-around section fitted with ConSep trays. For the remaining sections of the column, the following internals were suggested:

- Stripping section : Shell HiFi*** trays
- Wash section : Mellapak**** Plus 252Y packing
- All other sections : Shell CS trays

As this was the first application of Shell ConSep trays in this service, a detailed study was carried out to address the risks associated with this application and the mitigations were applied in the design. The ConSep trays were manufactured by Sulzer Chemtech. To ensure proper performance of the trays in a relatively new application, very rigorous quality control steps were followed at the manufacturing site and detailed mock-up assembly of tray components was carried out at refinery site prior to installation in the column (Figure-5).

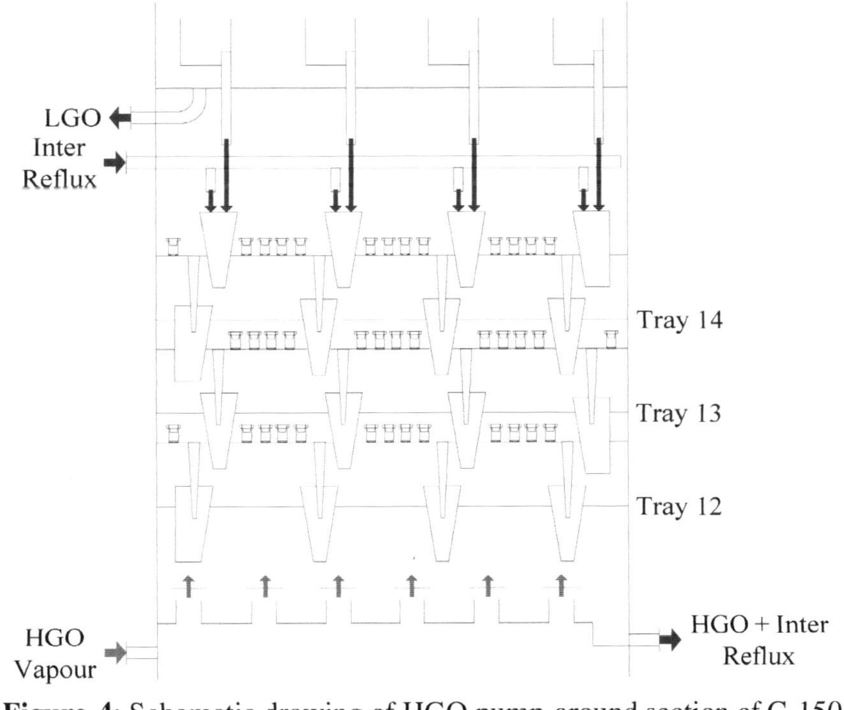

Figure-4: Schematic drawing of HGO pump-around section of C-150

Post Revamp Performance

From conception to implementation, the project took 4 years and the revamp was implemented during a normal planned shutdown in October 2009 without affecting the unit availability. The performance test run of CDU-1 was conducted in September 2010. The crude throughput was maintained at 13000 ton/day, though the crude blend used during the test run was marginally heavier than the one considered for the revamp design. Comparison of the design crude blend and crude blend used during test run is shown in Figure-6.

Figure-5: Inspection/Mockup/Installation of ConSep trays

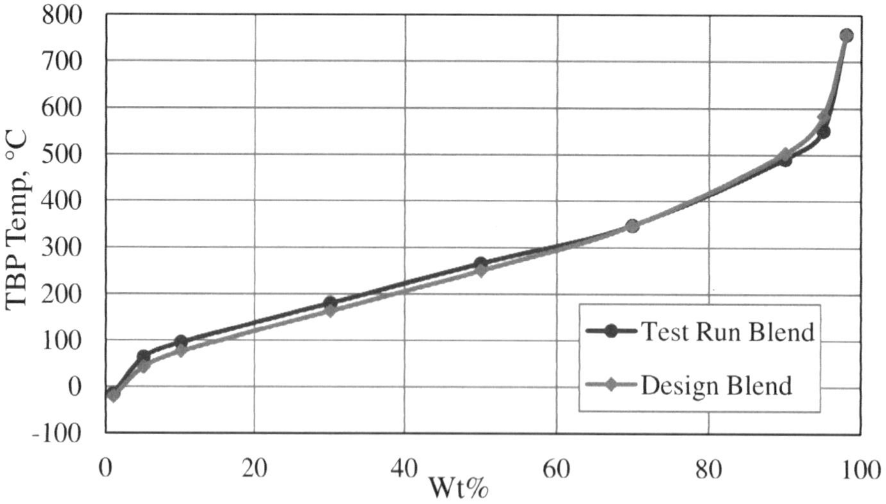

Figure-6: Comparison of design and test run crude TBP

Key design conditions and test run operating conditions of C-150 are presented in Table-2. The product quality is compared in Figure-7(a-e).

Table-2: Key design and test run operating conditions of C-150

Parameters	Unit	Design	Test Run	Parameters	Unit	Design	Test Run
Crude Intake	T/D	13000	13077	Flash Zone Press	barg	1.90	2.24
Naphtha	T/D	4405.0	3516.4	Feed Temp	°C	361.5	345.0
Kerosene	T/D	1498.8	2169.3	Top Temp	°C	180.8	175.7
Light Gas Oil	T/D	1517.0	1116.1	Kero Draw Temp	°C	216.5	223.4
Heavy Gas Oil	T/D	2014.5	1731.9	LGO Draw Temp	°C	244.2	250.1
Long Residue	T/D	3573.3	4271.1	HGO Draw Temp	°C	280.7	278.3
Strip Steam	T/D	106.9	149.5	Flash Zone Temp	°C	343.9	327.3
Top Press	barg	1.65	1.95	Bottom Temp	°C	333.3	322.061

C-150 was simulated for the test run conditions to evaluate hydraulic loading of the ConSep trays. In Table-3, the key performance indicators for the ConSep trays operating under test run conditions are compared with the design conditions. During the test Run, the ConSep trays were operating at 10-15% lower than the design capacity even at design intake of 13000 T/D due to:

- Heavier feed than the design case
- Less preheat recovery of the order of 10-12°C due to limitation in crude preheat train

7a: Naphtha ASTM D86

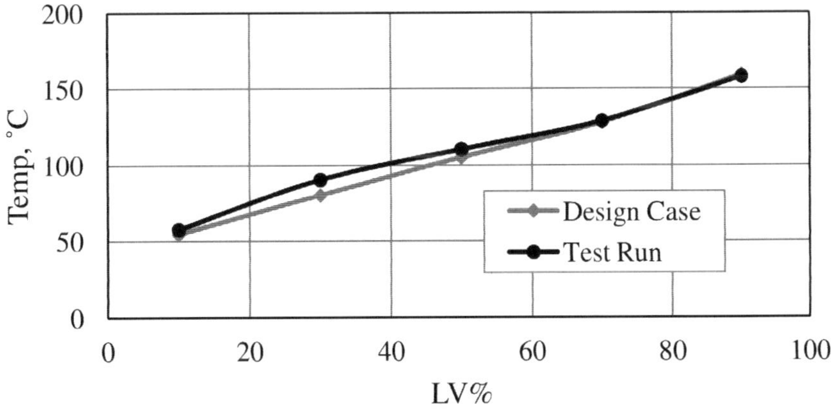

7b: Kerosene ASTM D86

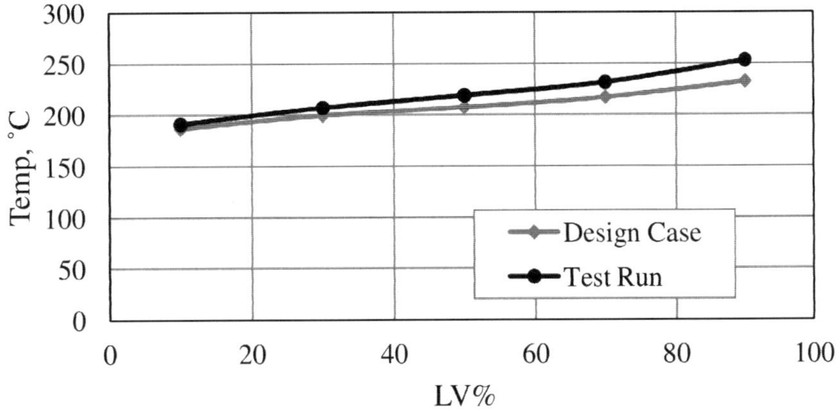

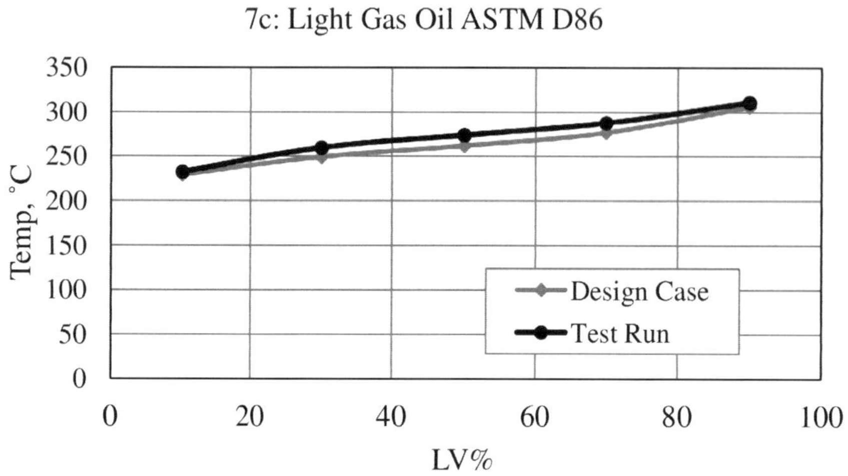

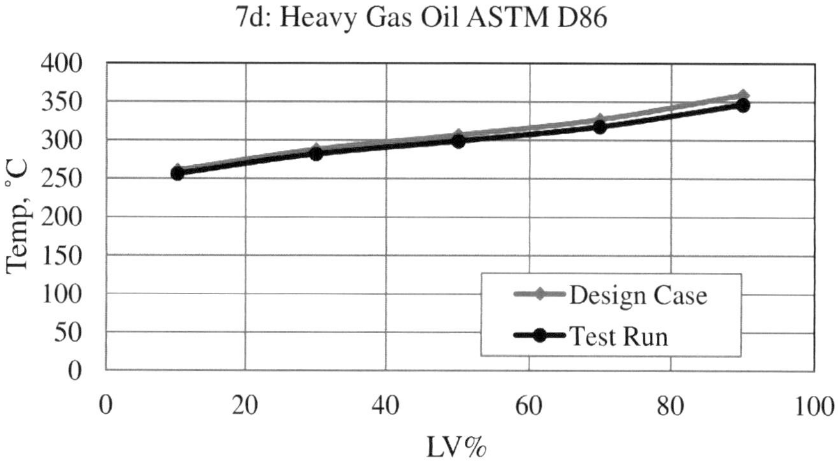

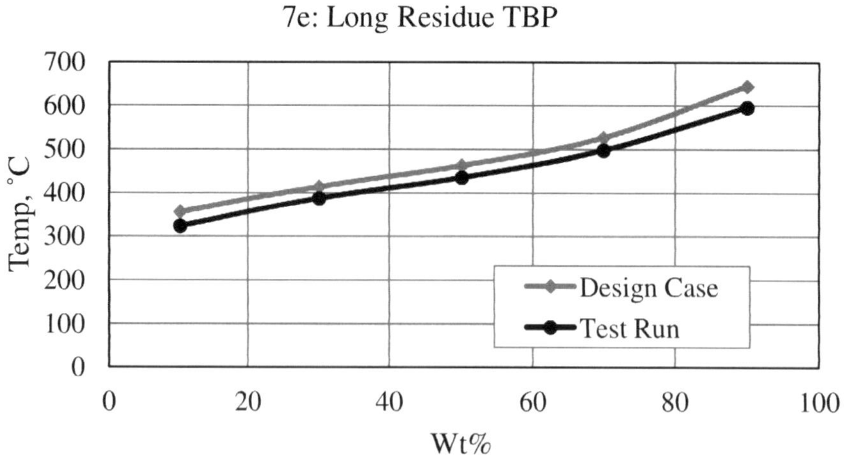

Figure-7(a-e): Comparison of design and test run product quality

Table-3: Key performance indicators for ConSep trays

Parameters	Unit	Design	Test Run
Froth backup / CS Height	%	68	60
Tray pressure drop	mbar	12.3	9.2
Tube flood percent	%	73	60
Flow Parameter	-	0.17	0.19
Overall column load factor	m/s	0.12	0.10
Flooding percent (CS tray)	%	133	112

Based on the review of the operating experience since start-up and the test run results, it could be concluded that the revamp targets of CDU-1 main fractionator (C-150) were achieved. No hydraulic constraint was experienced in achieving the design intake of 13000 ton/day and required product quality was achieved.

Conclusions

The performance of Shell ConSep trays in the HGO pump-around section of CDU-1 main fractionator met the target of capacity enhancement without any draw back compared to the pre-revamp conditions. During the test run, the trays were operating at 10-15% lower than the design capacity even at the design intake of 13000 T/D due to heavier crude feed and lower feed temperature. However, the built-in capacity margin enabled stable operation for the trays at much above the capacity limit of the first generation of high capacity trays.

The options to debottleneck columns already equipped with the first generation of high capacity trays are limited. Shell ConSep trays provide a very attractive solution to such cases. In this revamp project, use of only 3 Shell ConSep trays in the most capacity constrained section of the column made it possible to retrofit the existing column and turned the capex option more attractive over the other debottlenecking options.

References

1. 'Refinery expansion means NZ more self reliant' media release by NZRC dated 16 July 2010
2. Wilkinson, P.M., De Villiers, W.E., Mosca, G., Tonon, L., 'Achieve Challenging Targets in Propylene Yield using Ultra System Fractionation Trays', ERTC 2006.
3. De Villiers, W.E., Bravo, J.L., Wilkinson, P.M., Summers, D.R., 'Further advances in light hydrocarbon fractionation', PTQ Summer 2004

(*) Shell ConSep is a Shell trademark
(**) Shell CS a Shell trademark
(***) Shell HiFi a Shell trademark
(****) Mellapak is a Sulzer-Chemtech trademark

(94e) Flow Pattern Optimization of a Sieve Plate Extraction Column Using CFD Simulation and PIV Measurement

Changchun Duan[1], **Bing Wang**[2], Chunjiang Liu[1] and Xigang Yuan[1], (1)State Key Laboratory of Chemical Engineering, Tianjin University, (2)Chemical Engineering, New Jersey Institute of Technology, Newark, NJ

Present work deals with flow pattern optimization of the continuous phase in a sieve plate extraction column using both computational fluid dynamics (CFD) simulation and particle image velocimetry (PIV) measurement. Firstly, flow pattern of a sieve plate extraction column (SPEC) with traditional configuration was simulated and it was found that there was a very large reverse flow region between every two plates. Then, flow pattern was optimized with the objective of minimizing reverse flow area since the reverse flow can decrease the mass transfer efficiency. Step by step, internal structure of the column was modified to minimize the reverse flow area and with the modification, flow pattern of the continuous phase was optimized. The strategies consist of inclining downcomers, adding diversion baffles, slotting the downcomers and diversion baffles, adjusting the number and size of slots and fixing small baffles laterally at the top slots on downcomers. In the course of evolution, five generations (11 types) of SPEC with different internal configuration were proposed and flow patterns in these columns were investigated. Finally, an optimal flow pattern nearly no reverse flow region was obtained. PIV experiments of three types of columns in 11 were carried out and it was found that the results of simulations and experiments are in good agreement. Besides, the mass transfer performances of all columns were investigated by CFD simulation and it was found that mass transfer efficiency was improved as the flow pattern was optimized. The flow pattern optimization strategies proposed here can be extended to the design of other process equipment such as distillation and reaction.

Keywords: flow pattern optimization, extraction column, CFD, PIV

(94f) Pressure Drop on Gas Tangential Velocity in Rotating Zigzag Bed

Yumin Li, College of Chemical Engineering & Materials Science, zhejiang University of Technology, Hangzhou, China

The pressure drop across Higee is one of the most important characteristics, and the liquid flow gives rise to intriguing and unusual effects on the pressure drop. The pressure drop of the wet Higee can be substantially lower than that of the dry one. This trend is quite different from the conventional packed and trayed tower. And the unusual effects also occur in the rotating zigzag bed(RZB) as a novel type of Higee. The pressure drop of RZB was studied using two methods of pressure drop model based on the gas tangential velocity and empirical correlation. Experiments were carried out in a RZB of 214mm inner-diameter, 486mm outer-diameter and 104mm axial-height rotor with air-water system. The radial distribution of gas tangential velocity in the RZB rotor was obtained. And the liquid flow reduces the gas tangential velocity. The gas tangential velocity was correlated by an empirical equation, on which the centrifugal pressure drop is predicted by the pressure drop model. Thereby, the frictional and momentum pressure drop are also predicted by the model. As a result, the phenomenon that the wet rotor pressure drop is lower than the dry rotor pressure drop is reasonably explained. The variation of the overall pressure drop of RZB is similar to that of the rotor pressure drop, and it is correlated by an empirical equation with good agreement.

(94g) Design of Distillation Columns for Ternary Heterogeneous Azeotropic Mixtures by Geometrical Methods

Cesar Augusto Sánchez Correa[1], **Miguel Ángel Gómez García**[2] and Gerardo Rodríguez Niño[1], (1)Chemical and Environmental Engineering Department, National University of Colombia, Bogotá, Colombia, (2)Chemical Engineering Department, National University of Colombia, Manizales, Colombia

The distillation columns for the fractionation mixtures with regions of partial miscibility allowed the possibility of coexistence of two o more liquid phases inside or in an external decanter. This feature of the distillation columns with heterogeneous mixtures implies some advantages because of two issues associated with the decanter: a) enables cross a thermodynamic separatrix leading to products located in different distillation regions; b) when differences in solubility are notaries enables high purity distillation products in the section where the decanter is localized.

The basic design of a heterogeneous distillation column consists in developing the following tasks: 1) locate the product composition region and verify the separability condition; 2) investigate the conditions of minimum energy consumption (minimum reflux calculation); 3) locate the feeding stages and calculating the number of theoretical stages for a finite reflux (higher than the minimum). Shortcut methods used for the basic design include the knowledge associated with the following geometric elements [1, 2, and 3]: a) the*reversible distillation trajectories*(RDT); b) the distillation borders at infinite reflux; c) the*Pitchfork distillation boundary(PDB); d) the structure of vapor-liquid-liquid equilibrium.*

In this work new tools are present for the basic design of a heterogeneous distillation column. The target is: a) present a continuation method for tracking of the RDT associated with the feeds and the products; b) present a continuation technique for tracking of the PDB; c) use the RDT and the PDB in association with the other geometric elements for the exploring of the conceptual design problem. Three examples of design are illustrated: dehydration of acetic acid with n-amyl acetate, synthesis of separation sequence for mixtures of water + n-pentanol + n-amyl acetate and a review of the dehydration of ethanol with benzene.

We conclude that there are some common and complicated situations in the design of azeotropic heterogeneous distillations: 1) the minimum reflux can be controlled by a tangent pinch point; 2) the range of allowable values for the reflux is very narrow; 3) the minimum reflux not coincide with a infinite number of stages; 4) the admissible specifications in the condenser-decanter depend of the vapor line extension and localization of the thermodynamic separatrix.

Key Words: Heterogeneous azeotropic distillation; reversible distillation; Pitchfork distillation boundary; continuation methods

References

[1]. Lucia, A.; Amale, A. y Taylor, R. "Distillation Pinch Points and More".*Computers & Chemical Engineering*, vol. 32, 2008, pp. 1342-1364.

[2]. Petlyuk, F. B.; Danilov, R. Yu. y Serafimov, L. A. "Trees of Reversible Distillation Trajectories and Structure of Trajectory Bundles for Sections of Adiabatic Columns".*Theoretical Foundations of Chemical Engineering*, vol. 42, No. 6, 2008, pp. 795-804.

[3]. Królikowski, A. R.; Królikowski, L. J. y Wasylkiewicz, S, K. "Distillation profiles in ternary heterogeneous mixtures with distillation boundaries". Chemical Engineering Research and Design, vol 89, July 2011, pp. 879-893

(106a) Synthesis of PVA Based Membranes Modified by Benzimidazole Group for Direct Methanol Fuel Cells

Pratima Gajbhiye, Chemical Engineering, Indian Institute of Technology, Kanpur, State of Uttar Pradesh, India, Jayant K. Singh, Chemical Engineering, Indian Institute of Technology, Kanpur, Ka, India and Anil Kumar, Department of Chemical Engineering, Indian Institute of Technology Kanpur, Kanpur, India

Abstract

In recent years, there has been a great effort to look for alternate membrane material and Polybenzimidazole (PBI) membranes have been synthesized where the conducting moiety of this polymer is benzimidazolium groups. The monomers used to synthesize these polymers are bis conjugated amines and diacids are expensive. In this work, we have synthesized grafted polymers of PVA as membrane materials where the grafts are either Poly Acrylic Acid, Poly maleic anhydride (PMA) or Poly Methacrylic acid (PMAA) and these materials were denoted by PVA-g-PAA, PVA-g-PMA, and PVA-g-PMAA. We then synthesized C1, C3 & C4 compounds (formula given at the bottom of Table S1) having bis conjugated amines whereas compound C2 having only one conjugated amine group. One of the steps in synthesizing C1 compound is nitration of phenyl ring, where there is a possibility of formation of two isomers out of which only one of them has strictly bis conjugated amines and can be polymerized with Isophthalic acid giving the formation of PBI. The other byproduct formed is an impurity and is minimized by carrying out the nitration step at low temperature but this cannot be completely removed from the product which is a mixture. Table S1 presents the summary of our experimental results on characterization of membrane, In the C1 to C4 compounds and in some cases (membranes 9, 10, 11), Sulfocuccinic acid were added to the casting polymer solution as additional crosslinking agent in order to achieve higher conductivity. We added the polymer, thus formed, to the casting solution, we first compared the results of the Fuel cell formed by adding C1 (C3 and C4) compound alone and heating at 180°C for some time. Table S1 has two sets of results, one on characterization (columns 1, 2, 3,4,5,6, and 7) of membranes while the other on Fuel cell (columns 8 to 11). For the latter columns 10 and 11 gave results on characterization on Fuel cell while columns 8 and 9 gave results when the Fuel cell was run on control mode for higher efficiency.

Experiments on membrane pore sizes are given in column 1 which lies between 8.88 to 9.52nm and they all are ultra filtration type. The values of the pore size are mostly governed by the nature of the graft chains, e.g. membranes 1,3,4,7 and 12 consist of PAA grafts and were found to have similar pore sizes. Similarly membranes 2,5,8,9 were of the similar pore size because the PVA have the same grafts of PMA. The next important thing is to determine the wall charges of these membranes and they were estimated by experimentally determining the rejection of Chromic (VI) acid solution through these membranes. Subsequently, a space charge model was proposed for the flow of the solution through their pores and a semi analytical solution was derived for the overall problem of separation of chromic acid solution. We observed that the high surface charge was because of the benzimidazolium groups present on the surface and they ranged from minimum of 465 to maximum of 525mV. The thermal degradation (TGA) experiments were carried out which showed that, through the modification of the polymer by benzimidazole groups, lead to stable membrane and the degradation temperature varies from 150°C for 10% weight loss and for approximately 80% weight reduction, a minimum of 300°C to 700°C.

The twelve membranes reported in Table S1 were first characterized completely in terms of pore size, wall charge, conductivity, IEC and thermal degradation. In fuel cell applications, higher temperatures give large current density and it is observed that some of the membranes (rows 1, 2, 5, 6, 7, 8 and 12) could easily be operated upto 150°C without the membrane degrading. However in our case, we insisted that the membrane of direct methanol fuel cell were operated at 90°C for which most of the fuel cell results for Nafion 117 are available. The catalysts used on the anode and cathode side were Pt-Ru and Pt loaded on carbon respectively, which were synthesized by following procedure given in literature. We find that in each case, the power density and the current density were higher and for some of our results (membranes 1,6,12) were atleast two to three times higher than that of Nafion 117 as seen in column 10 and 11.

It was recognized that the membrane material has hydrophilic PVA matrix and hydrophobic benzimidazolium groups at the surface formed by reaction of the C1 to C4 compounds with grafts (see Fig.S-1). When the reaction occurs (confirmed by IR and NMR), the hydrophobic benzimidazolium are formed which tends to separate as confirmed by the SEM results. In view of this we assumed that the pores of the membrane would have a thin hydrophobic layer with positive charge walls. This is confirmed by the contact angle measurements where the minimum contact angle is 51° and varied upto 89°. Even though most of the polymer matrix is PVA with its grafts, when the ions move through the pores (giving rise to current in Fuel cell), the pore walls get highly charged (as measured by column 2 of Table S1). Some of the membranes have Sulfo succinic acid SSA as part of the matrix (e.g. the membrane 9 to 11) which would tend to give sulfonate groups inside the PVA matrix as well as at the phase boundaries, this way giving rise to bipolar nature of the membrane (benzimidazolium group being strictly cationic in nature). In addition to this, some of the compounds such as C1 and C2 compounds have bonded sulfonate groups which would give the positive and negative charge separated by the molecular distance, once again producing a bipolar nature to the membrane. In remainder of the cases the charges would be induced on the phase boundaries as seen in Fig.S-1. The conductivity observed in our membranes is very high and could be attributed to the special bipolar structure of our polymers.

Table S1. Fuel cell and Membrane Characterization

Compound	Membrane	1	2	3	4	5	6	7	8	9	10	11

			MWCO (pore size, nm)	Wall Charge	Conductivity (S/cm)	IEC	Thermal Degradation temp (°C) 10% - 80%		Water Uptake	Contact Angle	Average Voltage Gain (V)	Average time cycle (min)	Power Density (mW/ cm)	Current Density (mA/ cm)
C1	1	PVA-g-PAA-C1	9.1	488	0.04	2.7	199.732	268.70	7.5	72	0.523	55.75	50.47	74.88
	2	PVA-g-PMA-C1	9.35	465	1.23×10^{-2}	1.8	142.659	447.6	10.691	74.917	0.568	42.896	0.305	3.81
	3	PVA-g-PAA-modified	9.52	488	1.51×10^{-2}	2.12	115.123	706.53 (28.93%)	13.665	75.581	0.778	69.85	37.37	64.44
C2	4	PVA-g-PAA-C2	9.52	525	6.59×10^{-2}	2.38	119.558	691.93	13.4	51.2	0.704	39.83	23.0020	50.55
	5	PVA-g-PMA-C2	9.36	519	1.04×10^{-2}	2.17	181.59	562.348	14.9	50.24	0.77	161.93	34.72	62.11
	6	PVA-g-PMAA-C2	8.99	482	1.16×10^{-2}	2.18	201.055	472.568	13.8	51.8	0.6066	147.81	76.72	92.33
C3	7	PVA-g-PAA-C3	9.0	512	2.51×10^{-2}	2.23	204.908	453.03	8.624	70.368	0.6883	82.55	36.397	75
	8	PVA-g-PMA-C3	9.5	510	3.51×10^{-2}	2.27	166.819	505.293	13.513	62.944	0.726	37.557	36.11	64.381
	9	PVA-g-PAA-SSA-C3	9.36	497	1.71×10^{-2}	2.51	154.914	813.769 (24.32%)	10.5	54.8	0.8096	99.423	23.45	57.778
	10	PVA-g-PMA-SSA-C3	9.83	480	2.28×10^{-2}	2.46	115.12	428.713	8.5	76.2	0.663	224.22	28.561	58.358
	11	PVA-g-PMAA-SSA-C3	8.88	525	3.66×10^{-2}	1.85	139.772	681.715 (30.78%)	7.2	89.2	0.77	18.36	20.832	48.11
C4	12	PVA-g-PAA-C4	9.5	495	2.56×10^{-2}	1.62	189.87	451.86	8.927	59.132	0.704	74.19	41.1	71
	13	Nafion 117			0.1								21.01	48.33

Compound C1, [Bis-3-Amino-4-{3-(triethylammoniumsulfonato) phenylamino}-phenyl sulfone] hydrochloride

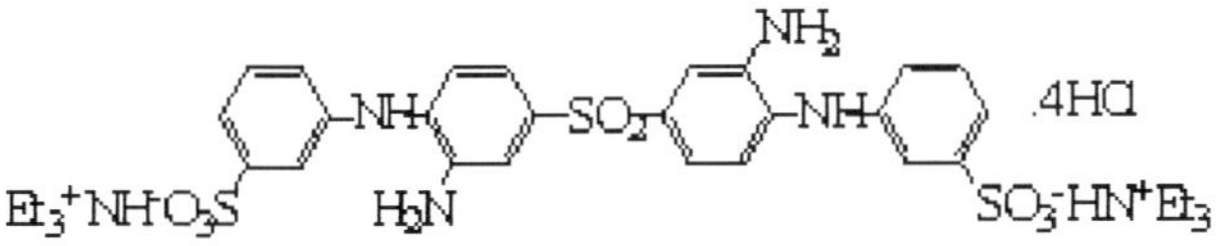

Compound C2, 3-Amino -4-[3-(triethyl ammonium sulfonato) phenyl amino] phenylene hydrochloride

THE GRAPHIC ON THIS PAGE WAS NOT AVAILABLE ON THE E-MEDIA VERSION OF THIS PAPER.

Compound C3, Bis -4 [1, 2 diamino phenyl]-phenyl sulfone) hydrochloride

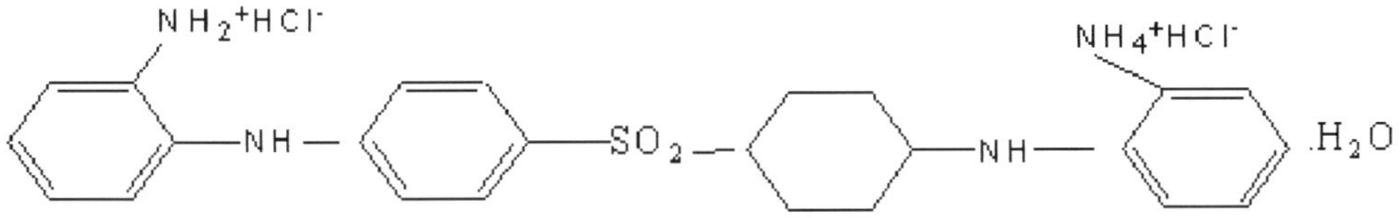

Compound C4, 1, 4 [1, 2 diamino phenyl] Phenylene hydrochloride

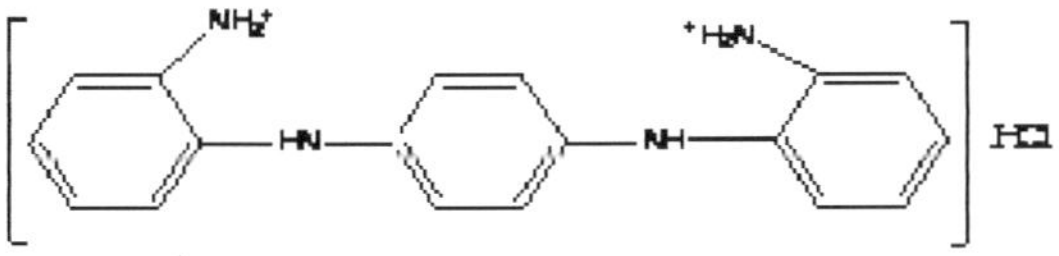

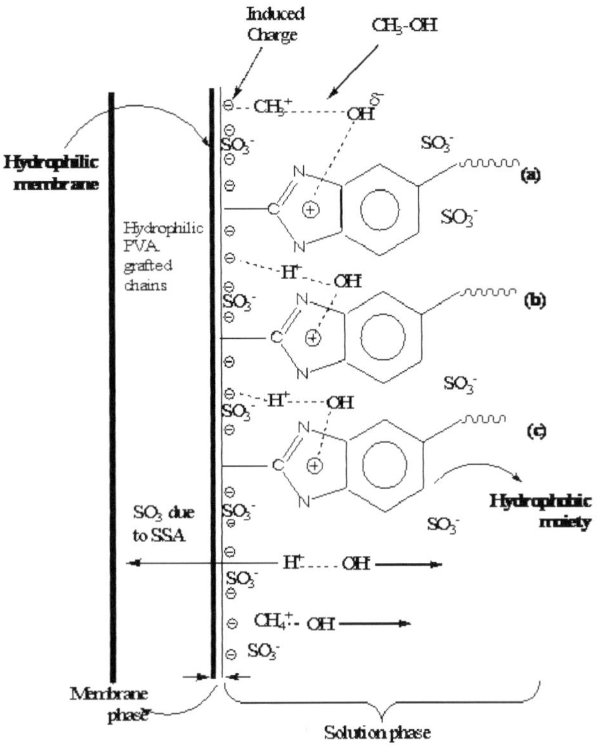

Figure S-1. The Bipolar nature of the membrane surface

(106b) Operation and Controls of Operating Direct Methanol Fuel Cells with PVA Based Membranes Modified by Benzimidazolium Groups

Pratima Gajbhiye, Chemical Engineering, Indian Institute of Technology, Kanpur, State of Uttar Pradesh, India, Pranjal Joshi, Chemical Engg, IIT Kanpur, Kanpur, India, Jayant K. Singh, Chemical Engineering, Indian Institute of Technology, Kanpur, Ka, India and A. Kumar, Department of Chemical Engineering, Indian Institute of Technology Kanpur, Kanpur, India

Operation and Control of Direct Methanol Fuel cell with PVA based membranes modified with benzimidazolium groups

Pratima Gajbhiye, Pranjal Joshi, Jayant K.Singh, Anil Kumar

Department of Chemical Engineering, Indian Institute of Technology, Kanpur 208016, India

Abstract

In recent years, in an effort to find alternate membrane material (other than Nafion 117), Polybenzimidazole (PBI) membranes have been synthesized where the conducting moiety of this polymer is benzimidazolium groups. The monomers used to synthesize these polymers are bis conjugated amines and diacids are expensive. In this work, we have synthesized grafted polymers of PVA as membrane materials where the grafts are either Poly Acrylic Acid, Poly maleic anhydride (PMA) or Poly Methacrylic acid (PMAA) and these materials were denoted by PVA-g-PAA, PVA-g-PMA, and PVA-g-PMAA. We then synthesized C1, C3 & C4 compounds (formula given at the bottom of Table S1) having bis conjugated amines whereas compound C2 having only one conjugated amine group. One of the steps in synthesizing C1 compound is nitration of phenyl ring, where there is a possibility of formation of two isomers out of which only one of them has strictly bis conjugated amines and can be polymerized with Isophthalic acid giving the formation of PBI. The other byproduct formed is an impurity and is minimized by carrying out the nitration step at low temperature but this cannot be completely removed from the product which is a mixture. Table S1 presents the summary of our experimental results on characterization of membrane, In the C1 to C4 compounds and in some cases (membranes 9, 10, 11), Sulfocuccinic acid were added to the casting polymer solution as additional crosslinking agent in order to achieve higher conductivity. We added the polymer, thus formed, to the casting solution, we first compared the results of the Fuel cell formed by adding C1 (C3 and C4) compound alone and heating at 180°C for some time. Table S1 has two sets of results, one on characterization (columns 1, 2, 3,4,5,6, and 7) of membranes while the other on Fuel cell (columns 8 to 11). For the latter columns 10 and 11 gave results on characterization on Fuel cell while columns 8 and 9 gave results when the Fuel cell was run on control mode for higher efficiency.

Experiments on membrane pore sizes are given in column 1 which lies between 8.88 to 9.52nm and they all are ultra filtration type. The values of the pore size are mostly governed by the nature of the graft chains, e.g. membranes 1,3,4,7 and 12 consist of PAA grafts and were found to have similar pore sizes. Similarly membranes 2,5,8,9 were of the similar pore size because the PVA have the same grafts of PMA. The next important thing is to determine the wall charges of these membranes and they were estimated by experimentally determining the rejection of Chromic (VI) acid solution through these membranes. Subsequently, a space charge model was proposed for the flow of the solution through their pores and a semi analytical solution was derived for the overall problem of separation of chromic acid solution. We observed that the high surface charge was because of the benzimidazolium groups present on the surface and they ranged from minimum of 465 to maximum of 525mV. The thermal degradation (TGA) experiments were carried out which showed that, through the modification of the polymer by benzimidazole groups, lead to stable membrane and the degradation temperature varies from 150°C for 10% weight loss and for approximately 80% weight reduction, a minimum of 300°C to 700°C.

The twelve membranes reported in Table S1 were first characterized completely in terms of pore size, wall charge, conductivity, IEC and thermal degradation. In fuel cell applications, higher temperatures give large current density and it is observed that some of the membranes (rows 1, 2, 5, 6, 7, 8 and 12) could easily be operated upto 150°C without the membrane degrading. However in our case, we insisted that the membrane of direct methanol fuel cell were operated at 90°C for which most of the fuel cell results for Nafion 117 are available. The catalysts used on the anode and cathode side were Pt-Ru and Pt loaded on carbon respectively, which were synthesized by following procedure given in literature. We find that in each case, the power density and the current density were higher and for some of our results (membranes 1,6,12) were atleast two to three times higher than that of Nafion 117 as seen in column 10 and 11. In normal Fuel cell operations, there is decay in the cathode catalyst which is further increased by the methanol crossover. The voltage of the Fuel cell increases, the average volatage gained is reported in column 8 and it is found that the voltage gain depends on the nature of the membrane which is not the same for all of them. In column 9, we have measured the average time cycle and we found that some of the membranes (membranes 6, 7, 9, 10) gave extremely good results.

It was recognized that the membrane material has hydrophilic PVA matrix and hydrophobic benzimidazolium groups at the surface formed by reaction of the C1 to C4 compounds with grafts (see Fig.S-1). When the reaction occurs (confirmed by IR and NMR), the hydrophobic benzimidazolium are formed which tends to separate as confirmed by the SEM results. In view of this we assumed that the pores of the membrane would have a thin hydrophobic layer with positive charge walls. This is confirmed by the contact angle measurements where the minimum contact angle is 51° and varied upto 89°. Even though most of the polymer matrix is PVA with its grafts, when the ions move through the pores (giving rise to current in Fuel cell), the pore walls get highly charged (as measured by column 2 of Table S1). Some of the membranes have Sulfo succinic acid SSA as part of the matrix (e.g. the membrane 9 to 11) which would tend to give sulfonate groups inside the PVA matrix as well as at the phase boundaries, this way giving rise to bipolar nature of the membrane (benzimidazolium group being strictly cationic in nature). In addition to this, some of the compounds such as C1 and C2 compounds have bonded sulfonate groups which would give the positive and negative charge separated by the molecular distance, once again producing a bipolar nature to the membrane. In remainder of the cases the charges would be induced on the phase boundaries as seen in Fig.S-1. The conductivity observed in our membranes is very high and could be attributed to the special bipolar structure of our polymers.

Table S1. Fuel cell and Membrane Characterization

Compound	**Membrane**		**1**	**2**	**3**	**4**	**5**		**6**	**7**	**8**	**9**	**10**	**11**
			MWCO (pore size, nm)	**Wall Charge**	**Conductivity (S/cm)**	**IEC**	**Thermal Degradation temp (°C) 10% - 80%**		**Water Uptake**	**Contact Angle**	**Average Voltage Gain (V)**	**Average time cycle (min)**	**Power Density (mW/ cm)**	**Current Density (mA/ cm)**
C1	1	PVA-g-PAA-C1	9.1	488	0.04	2.7	199.732	268.70	7.5	72	0.523	55.75	50.47	74.88
	2	PVA-g-PMA-C1	9.35	465	1.23×10^{-2}	1.8	142.659	447.6	10.691	74.917	0.568	42.896	0.305	3.81
	3	PVA-g-PAA-modified	9.52	488	1.51×10^{-2}	2.12	115.123	706.53 (28.93%)	13.665	75.581	0.778	69.85	37.37	64.44
C2	4	PVA-g-PAA-C2	9.52	525	6.59×10^{-2}	2.38	119.558	691.93	13.4	51.2	0.704	39.83	23.0020	50.55
	5	PVA-g-PMA-C2	9.36	519	1.04×10^{-2}	2.17	181.59	562.348	14.9	50.24	0.77	161.93	34.72	62.11
	6	PVA-g-PMAA-C2	8.99	482	1.16×10^{-2}	2.18	201.055	472.568	13.8	51.8	0.6066	147.81	76.72	92.33
C3	7	PVA-g-PAA-C3	9.0	512	2.51×10^{-2}	2.23	204.908	453.03	8.624	70.368	0.6883	82.55	36.397	75
	8	PVA-g-PMA-C3	9.5	510	3.51×10^{-2}	2.27	166.819	505.293	13.513	62.944	0.726	37.557	36.11	64.381
	9	PVA-g-PAA-SSA-C3	9.36	497	1.71×10^{-2}	2.51	154.914	813.769 (24.32%)	10.5	54.8	0.8096	99.423	23.45	57.778
	10	PVA-g-PMA-SSA-C3	9.83	480	2.28×10^{-2}	2.46	115.12	428.713	8.5	76.2	0.663	224.22	28.561	58.358
	11	PVA-g-PMAA-	8.88	525	3.66×10^{-2}	1.85	139.772	681.715	7.2	89.2	0.77	18.36	20.832	48.11

		SSA-C3						(30.78%)						
C4	12	PVA-g-PAA-C4	9.5	495	2.56×10^{-2}	1.62	189.87	451.86	8.927	59.132	0.704	74.19	41.1	71
	13	Nafion 117			0.1								21.01	48.33

Compound C1, [Bis-3-Amino-4-{3-(triethylammoniumsulfonato) phenylamino}-phenyl sulfone] hydrochloride

Compound C2, 3-Amino -4-[3-(triethyl ammonium sulfonato) phenyl amino] phenylene hydrochloride

Compound C3, Bis -4 [1, 2 diamino phenyl]-phenyl sulfone) hydrochloride

Compound C4, 1, 4 [1, 2 diamino phenyl] Phenylene hydrochloride

THE GRAPHICS ON THIS PAGE WERE NOT AVAILABLE ON THE E-MEDIA VERSION OF THIS PAPER.

Figure S-1. The bipolar nature of PVA-g-PAA-BASPAPS membrane which reduces the methanol crossover due to the molecular dissociation. In (a), the imidazolium group is shown to separate out from the PVA moiety of the chain into a separate phase. In (b), the water is shown to complex with the SO and the imidazolium group. As a result of this, interaction in (d) and (e), the water and the methanol is shown to split into anion and cation and the anion permeates through the membrane giving bipolar effect

(106e) Numerical Investigation in Heat Conduction Effect on Heat Exchange for Mini Cross-Flow Heat Exchanger (MCHE)

Hongyue Duan[1], Wei Mu[1], **Xigang Yuan**[1] and Lingai Luo[2], (1)State Key Laboratory of Chemical Engineering, Tianjin University, Tianjin, China, (2)LOCIE Chambery, Savoie University, Savoie, France

In a traditional heat exchanger, the effect of heat transfer within the wall material driven by the temperature gradient in axial direction of the fluid flow is usually very small because the area of the cross section of the tube wall is much smaller than that of the tube for heat exchange. However, in a mini cross-flow heat exchanger(MCHE), where the heat conducting medium is typically a cubic-shaped block of metal and the hot and cold streams are put into heat exchange via two sets of channels in the block, the heat conduction in the wall material in the direction parallel to the fluid flow may have a significant effect on the performance of the heat exchanger.

In this paper, a 2-channel-layer MCHE was simulated and a periodic boundary condition was applied so that the results could be applicable to entire MCHE. The proposed model and the mesh grid size were proved by the good agreement of the simulated overall heat transfer coefficient with that found in the experimental reported in the literature[1]. Then the effect of heat conduction on heat transfer performance in a MCHE was studied numerically. The simulation results suggest that the performance of the MCHE can be improved by increasing the wall thickness between the adjoining channels in the same layer or by decreasing the wall thickness between the adjacent layers. The heat conduction effect can also be weakened by optimizing the material conductivity or by increasing the diameter of the channels. New correlations were developed based on the simulated results to predict the value of the overall heat transfer coefficient as a function of the operating parameter such as Re in the channel and the design parameters of the MCHE such as the diameter of the channel, wall thickness and thermal conductivity.

[1]Luo L A, Fan Y L, Zhang W W*et al.*Integration of constructal distributors to a mini crossflow heat exchanger and their assembly configuration optimization[J]. Chemical Engineering Science, 2007, 62 (13): 3605-3619.

(106g) The Effects of Out Gassing Through Cylindrical Walls on the Transport Coefficients

Luma. H. Yaseen[1],**Nada B. Nakkash**[2]and M. N. Latif[1], (1)Al-Nahrain University, Baghdad, Iraq, (2)Chemical Engineering, Al-Nahrain University, Baghdad, Iraq

Abstract

This work concerns with studying the influence of blowing on friction factor and mass transfer coefficient when the flow is normal to a tube bank with outgassing porous cylinders. The friction factor and the mass transfer coefficients are measured for a range of Reynold's number and blowing rate and expressed in terms of various dimensionless groups.

The test rig consists of a rectangular duct (20.3 × 11.3 cm), five rows of three porous tubes in staggered arrangements fixed across the duct with 1.5pitches (both longitudinal and transversal). The outside and the inside diameters of tubes were 25.4 and 21.4 mm, respectively.

The friction factor measurements was studied for Reynold's number range of

$3000 < Re < 17{,}000$and blowing rate up to 4×10^{-3}. The results show that the blowing intothe boundary layer for the cylinders increases the overall friction factor from 10%to 50% depending on the blowing rate. The mass transfer coefficient was measured for Reynold's number range of $5000 < Re < 16{,}000$ and blowing rate of $10^{-5} < m/G_\infty < 10^{-3}$.

A new definition of the blowing rate was considered and expressed by the mass transfer driving force. Two successful formulas were obtained to account the outgassing effect on the mass transfer coefficient. The mass transfer coefficient for the outgassing boundary layer was obtained first by determining the non-blowing coefficient and multiplying it with the correction factor .

The first model was modified to account the density variation associated with the boundary layer change:

$$St_m Sc^{2/3} = \frac{0.51\, Re^{-0.5}}{(1+\beta_m)^{4.7}} \left(\frac{\rho_s}{\rho_\infty}\right)^{0.4}$$

With the coefficient of determination R2 = 0.93.

The second model depends on the flat plate correction factor:

$$St_m Sc^{2/3} = 0.5\, Re^{-0.5} \left(\frac{Ln(1+\beta_m)}{\beta_m}\right)^{9.94}$$

With the coefficient of determination R2 = 0.92.

These two models are found to be in agreement with the previous literature and a reduction of about 80% in the mass transfer coefficient has been obtained due to the outgassing.

(106h) Down the Hubbert Curve

Ian Sutton, Sutton Technical Books, Houston, TX

Down the Hubbert Curve

In March 1956 M. King Hubbert, who was at that time Chief Geologist with the Shell Development Company published a seminal paper with the title "Nuclear Energy and the Fossil Fuels" (Hubbert 1956). In this lengthy and authoritative paper, which was presented before the American Petroleum Institute in San Antonio, Texas, Dr. Hubbert showed how key fossil fuel resources in the United States, particularly coal and oil, were of finite size and that their output would reach a maximum, and then decline. Dr. King's forecasts for the production of oil in the United States were accurate; he predicted that oil production would peak in the early 1970s, and that is what has happened. (The widely-used term "Peak Oil" is frequently accredited to Dr. Colin Campbell (2001)).

In the fifty years since the publication of Dr. King's paper great advances have been made in finding and extracting oil from what were considered to be marginal fields. But Dr. King's overall conclusion is still valid. Production of oil in the United States remains below the 1970's peak. Although Dr. Hubbert's analysis was confined to the United States, subsequent analysts applied his insights to other parts of the world. For example, the book*Twilight in the Desert: The Coming Saudi Oil Shock and the World Economy*(Simmons 2005) by the late Matthew Simmons achieved wide recognition by showing that the concept of peak oil most likely applies to reserves in Saudi Arabia, and other nations such as Iran and Kuwait.

The purpose of this paper is to examine how oil production and consumption have changed in the last five years, and to examine whether or not the concept of Peak Oil remains valid. Key to this discussion is an understanding of the concept of Energy Returned on Energy Invested (EROEI). It is concluded that the concept of Peak Oil, as first postulated by Dr. King, remains valid and current, and the production of oil peaked in the period 2005-2011 and is now set to move down the right hand side of the Hubbert Curve.

The title of this session is*Challenges in Instituting Energy Savings Project in the Present Market Conditions*. If the production of oil and has indeed already passed a peak then the challenge is not to do with "instituting" energy savings; those savings are going to occur whether we like it or not. The challenge will be to determine how the process industries can restructure themselves given that the energy resources available (both as direct energy and as petrochemical feedstock) will be in decline and that there will be intense demand for those same resources from other areas of the economy.

(106j) Statistical Decision Making for In-Service Inspection of Underground Waste Storage Tanks

Emily Mitchell, Emory University, Atlanta, GA and **Stephen P. Harris**, Savannah River National Laboratory, Aiken, SC

Pitting is a localized form of corrosion that occurs on passivated metal surfaces exposed to aggressive chemical environments. The pitting process involves the breakdown of the passive oxide film and subsequent anodic dissolution of the metal at a local site, while the remainder of the surface remains protected by the film.

Deep pits may result in lowering the maximum level of waste that made be stored in the tank and eventually lead to loss of confinement. Pit depths, established by ultrasonic testing, are subject to variables that cause uncertainty, including the repeatability of metric location, time between measurements, different transducers, and different inspectors.

The generalized extreme value (GEV) distribution is fit to the data gathered from ultrasonic testing and statistical projections are made for the full tank wall interior. A comparison is made with results from other well known statistical distributions. The estimated parameters are used in statistical tolerance limits for bounding pit depth in a waste tank.

The tolerance limits give an indication of extreme values of pit depth for the current condition of the tank walls. The upper tolerance limit for pit depth will then be compared to action levels for pitting. If the upper tolerance limit for pit depth exceeds the action level, more frequent examinations or increased tank surface area inspections are conducted.

(106k) Thermal Inbreathing Requirements of Low Pressure Storage Tanks At Elevated Temperatures

Donald E. Brooks, Pressure Relief Analysis, Lloyd's Register Energy Americas, Inc., Houston, TX

ISO 28300 (API 2000) "Petroleum, petrochemical and natural gas industries - Venting of atmospheric and low pressure storage tanks" provides guidance on determining the required thermal inbreathing flow rates to protect atmospheric and low pressure storage tanks against vacuum due to weather changes (e.g. a decrease in atmospheric temperature)." For heated, un-insulated tanks with vapor space temperatures above 58.9 °C (120 "F), ISO 28300 recommends an engineering review of thermal inbreathing requirements.

This paper describes one approach to evaluating thermal inbreathing requirements for un-insulated storage tanks with vapor space temperatures exceeding 58.8 °C (120 "F)." The tank is assumed to be exposed to the cooling effects of a sudden rain. A simplified unsteady state heat transfer analysis is performed to determine the tank wall and vapor space temperatures versus time. Initial gas mass and pressure are assumed to remain constant in the tank. Gas volume in the tank is calculated as a function of vapor space temperature using the ideal gas law. The rate of volume change is calculated for each time increment. The rate of volume change goes through a maximum value. This maximum value is selected as the thermal inbreathing requirement of the tank.

The calculation methodology described in this paper can be programmed in a spreadsheet for ease of use.

Dynamic Operational Risk Assessment with Bayesian Networks

Shubharthi Barua, Xiaodan Gao, M. Sam Mannan
Mary Kay O'Connor Process Safety Center
Texas A&M University, Jack E. Brown Engineering Bldg., 3122 TAMU, Room 244,
College Station, TX 77843
shubharthi.barua@tamu.edu, xiaodan_gao08@mail.che.tamu.edu, mannan@tamu.edu

Extended Abstract Prepared for the Poster Presentation at the
8th Global Congress on Process Safety
Houston, TX
April 1 – 4, 2012

UNPUBLISHED

8TH GLOBAL CONGRESS ON PROCESS SAFETY

Dynamic Operational Risk Assessment with Bayesian Networks

Shubharthi Barua, Xiaodan Gao, M. Sam Mannan

Mary Kay O'Connor Process Safety Center
Texas A&M University, Jack E. Brown Engineering Bldg., 3122 TAMU Room 244, College Station, TX-77843
shubharthi.barua@tamu.edu, xiaodan_gao08@mail.che.tamu.edu, mannan@tamu.edu

Keywords: risk assessment, Bayesian network, dynamic Bayesian network, fault tree, event tree

Abstract

Oil/gas and petrochemical plants are complicated and dynamic in nature. Dynamic characteristics include ageing of equipment/components, season changes, stochastic processes, operator response times, inspection and testing time intervals, and timing of safety system operations, all of which are time dependent criteria that can influence dynamic processes. Conventional risk assessment methodologies can quantify dynamic changes in processes with limited capacity. Several methods are proposed for dynamic risk assessment such as C-DET, D-DET, DORA, etc., but all of them have deficiencies of some kind. Bayesian networks, a graphical inference technique, have some advantageous characteristics for risk analysis over other techniques. This research focuses on providing a framework for dynamic risk analysis based on Bayesian networks. A case study will be provided to illustrate this model.

1.0 Introduction

Oil/gas, chemical, and petrochemical process plants consist of numerous equipment and unit operations, thousands of control loops, and exhibit dynamic behavior. Different hazards and several types of risk are associated with these process facilities. For example, large process areas have the potential to stock a large amount of hazardous chemicals, which can result in the process area being highly congested. It is reported that process risk, i.e., risk in process facilities due to fires and explosions, greatly contributes to the total risk of offshore installations. Also, reactivity, toxic, and mechanical hazards have a significant effect on the total risk of these facilities. Therefore, it is very important to identify hazards, perform risk assessments, and take proper initiatives to minimize/remove hazards and risks; else a catastrophic accident may result.

From case histories, it has been observed that catastrophic accidents have a significant effect on people, environment, and society. Catastrophic accidents such as the Flixborough disaster, the Bhopal incident, and the Piper Alpha disaster caused fatalities and unbearable economic loss. In the past few years, several disastrous accidents in refineries, power plants, and offshore platforms have significantly affected people's perception, and consequently raised concern to emphasize process safety. A series of explosions in 2005 resulting from insufficient process safety and lack of risk reduction measures at the BP Texas City refinery accident resulted in 15 fatalities, 180 injuries, and $1.5 billion in losses [1]. The Kleen Energy power plant explosion in

Connecticut in 2010 is considered to be one of the worst industrial explosions in the USA [2]. In 2010, a fire and explosion resulting from a blowout at the Macondo well resulted in 11 deaths and 17 injuries. Also the continuous spill from the wellhead for 87 days had disastrous effects on the environment and wildlife surrounding the Gulf of Mexico. It has been reported that most of the accidents could be prevented or their severity could have been minimized by effective risk assessment and process safety management. But existing risk assessment methodologies cannot properly quantify various time-dependent effects. Therefore, the objective of this research is to develop a risk assessment methodology with the capability to capture variations due to different time-dependent effects.

2.0 Literature Review

N. Siu [3] summarized available methodologies for risk assessment. For example, the fault tree and event tree methods are two most common methodologies for risk assessment. But both techniques have limited ability to address different time-dependent effects such as equipment or component ageing, stochastic processes, operator response times, and inspection and testing time intervals. N. Siu [3] also described other methodology limitations while applying dynamic system analyses. Therefore, for dynamic systems, conventional techniques are not able to provide accurate results because of their inadequate ability to describe the variation of operational risks as time-dependent deviations, or the changes occurring in the process. Continuous event trees [4], discrete event trees [5], and Markov theory [6] methods have been proposed for dynamic reliability and safety analysis. But all of the proposed techniques have some deficiencies. For example, the Markov theory based model does not consider the effect of inspection on system-state transitions. X. Yang et al. [1] proposed the Dynamic Operational Risk Assessment (DORA) technique, which is able to take inspection and testing time intervals and repair time into account for evaluating operational failure.

Recently, Bayesian networks are receiving wide acceptance in reliability assessment, fault diagnosis, safety, risk, and maintenance analysis. Several authors have [7,8] compared some commonly used methods for accident analysis with Bayesian networks, and described several advantages of Bayesian networks over these methods. N. Khakzad et al. [2] illustrated parallelism between Bayesian networks and fault trees, and discussed the potential of Bayesian networks as a tool for process safety and accident analysis. H. Pasman and W. Rogers [9] described an approach to incorporate Bayesian networks into layer of protection analysis to make it more effective. These studies demonstrate the potential advantages of Bayesian networks application in risk assessment. This research provides a methodology for dynamic risk assessment based on Bayesian networks.

3.0 Methodology and Expected Result

3.01 Bayesian Networks

A Bayesian network, which is a probabilistic graphical model, describes causal influence relations among variables. A set of random variables and their conditional dependencies are represented by Bayesian networks via a directed acyclic graph (DAG). It has the ability to represent dependency among events clearly, accommodate multi mode and continuous random variables, and incorporate information i.e. generic, system specific and expert judgment to support optimum decision making.

A simple Bayesian network is shown in Figure 1. In a binary network, variables are represented by nodes and causal relationships among different nodes are shown by an arc. Conditional probability tables or defined probabilistic relations among nodes represent how one variable is linked with one or many variables. The nodes that influence another variable and have unconditional probability are called parent or root nodes. Nodes that are conditionally dependent on their direct parents are called intermediate nodes. The top node is defined as a leaf node, which is similar to the top event of a fault tree and initiating event for an event tree.

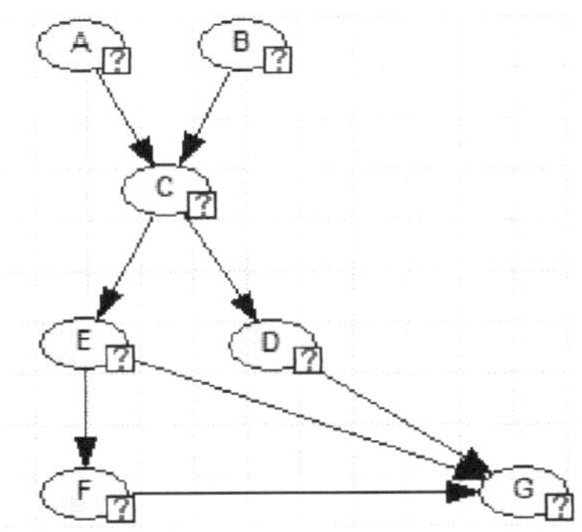

Figure 1: A simple Bayesian network

Let $N = (G, P)$ be a Bayesian network, where

$G = (V, E)$ is a directed acyclic graph; V (random variables) represents nodes; and E represents edges between pairs of nodes of DAG.

P represents probability distribution over V and $V = \{X_1, X_2, \ldots, X_n\}$ is either discrete or continuous random variables [10]. These random variables are assigned to the nodes and the edges.

Bayesian networks can be represented by the joint probability distribution P(V);

$$P(V) = \sum_{X \in V} P\{X|pa(X)\} = P(X_1, X_2, \ldots, X_N) = \sum_{i=1}^{n} P\{X_i|pa(X_i)\}$$

Here $pa(X_i)$ = parent nodes of X_i.

A main advantage of the application of Bayesian networks in risk analysis is the ability to update prior data using Bayes' theorem by incorporating new information as demonstrated by N. Khakzad et al. [2]. Also, Bayesian networks have an advantage of handling different types of uncertainty.

3.02 Dynamic Bayesian Networks

A general Bayesian network is a static model, i.e., the joint probability distribution is usually a representation of a fixed point or an interval of time [11]. A dynamic Bayesian network is a Bayesian network that has the ability to describe the evolution of joint probability distribution over time. Dean and Kanazawa [12] proposed modeling time in a discrete manner to represent the progression of time in a dynamic Bayesian network. In a dynamic Bayesian network, arcs links node/nodes from previous time slice to the node/nodes of the next time slice to represent temporal dependencies among them.

3.03 Methodology

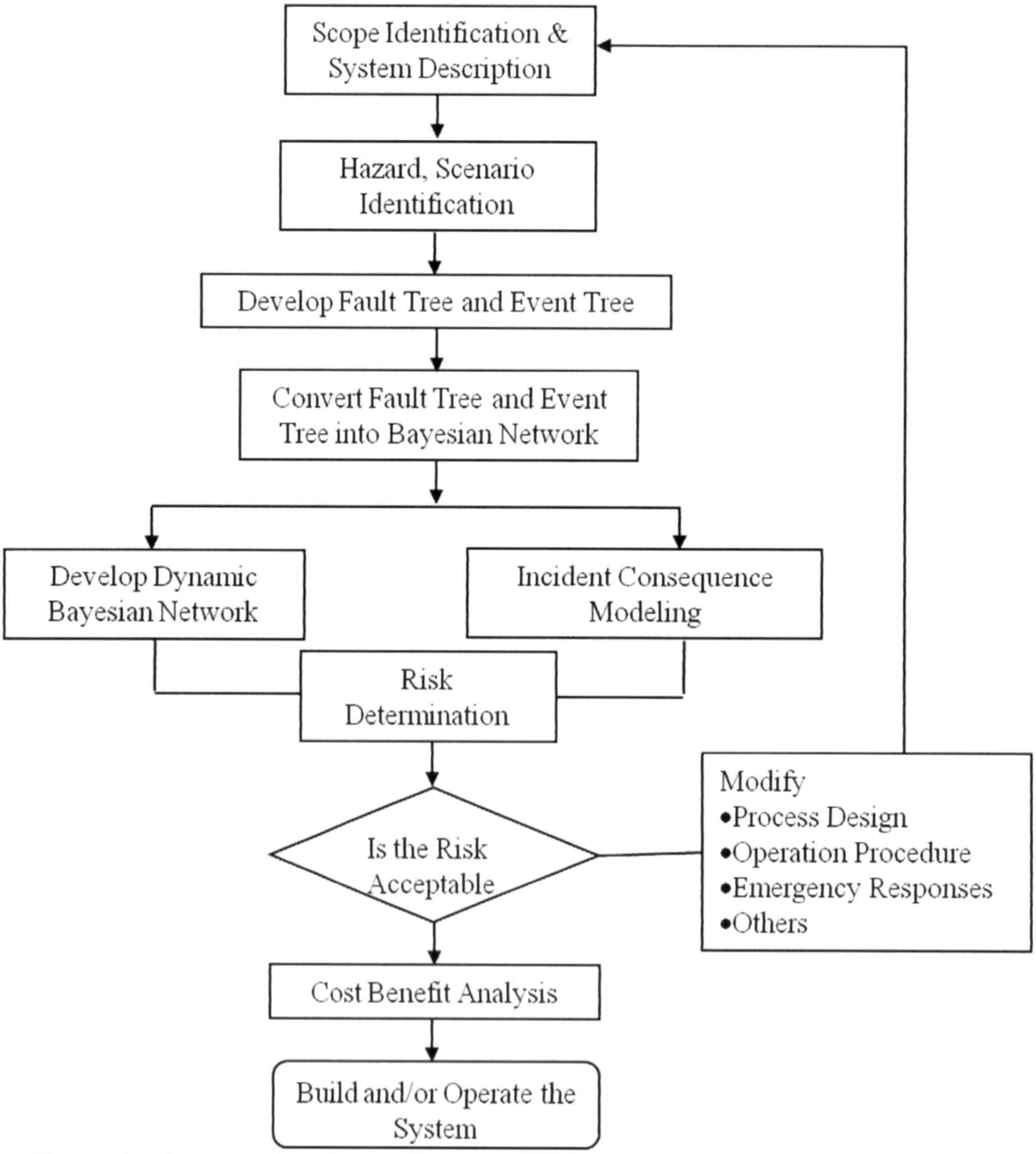

Figure 2: Framework for the dynamic Bayesian network based risk assessment

The objective of this research is to propose a dynamic risk assessment method based on dynamic Bayesian networks. According to the framework, a fault tree and event tree are developed for a system. The objective for following step is to convert the developed fault and event tree into a Bayesian network. Bobbio et al. [13] provided a methodology for conversion of fault trees into Bayesian networks. N. Khakzad et al. [2] showed equivalency between fault trees and Bayesian networks, and described the advantages of Bayesian networks for process safety analysis. Bearfield and March [14] proposed another methodology for generalizing event trees into Bayesian networks. The method has the ability to generalize simple event trees into Bayesian networks in a relatively easy manner, but the method becomes complicated for large event trees. Therefore, in the first phase, a methodology is proposed to convert fault trees and event trees into a static Bayesian network. In the second phase, a dynamic Bayesian network is developed from the static Bayesian network. The dynamic Bayesian network provides probability of a system at different time-slices for different states of the system. The probabilities are then combined with consequence modeling levels to calculate risk. GeNIe, which was developed at the Decision Systems Laboratory at the University of Pittsburgh, is used to create the model [15].

4.0 Conclusions

This article briefly describes a framework based on Bayesian network for dynamic risk assessment for the oil/gas, chemical, and petrochemical industries. This research illustrates the capability of Bayesian networks to represent all component interactions in complex systems, and to update all prior system probabilities in the presence of new information. The advantages of modeling flexibility and efficient computational mechanisms using dynamic Bayesian networks are also discussed.

4. References

[1] Yang, X. and Mannan, M.S., "The development and application of dynamic operational risk assessment in oil/gas and chemical process industry", Reliability Engineering and System Safety 95 (2010) 806-815

[2] Khakzad, N., Khan, F. and Amyotte,P., "Safety analysis in process facilities: Comparison of fault tree and Bayesian network approaches", Reliability Engineering and System Safety 96(2011) 925-932

[3] Siu,N. "Risk assessment for dynamic systems: an overview", Reliability Engineering and System Safety 43 (1994) 43-73

[4] Devooght, J. and Smidts,C. "Probabilistic reactor dynamics-I: the theory of continuous event trees", Nuclear Science and Engineering 1992; 111:229-40

[5] Amendola, A. "Accident sequence dynamic simulation versus event trees", Reliability Engineering and System Safety 1988; 22:3-25

[6] Howard, R. "Dynamic probability systems", Vols. I and II, New York, 1971

[7] Sklet, S. "Comparison of some selected methods for accident investigation", Journal of Hazardous Materials 2004; 111:29-37

[8] Nivolianitou, Z.S., Leopoulos, V.N. and Konstantinidou, M., "Comparison of techniques for accident scenario analysis in hazardous systems", Journal of Loss Prevention in the Process Industries 2004; 17"467-75

[9] Pasman, H. and Rogers, W.; "BBN, a tool to make LOPA more effective, QRA more transparent and flexible, and therefore to make safety more definable", 14th Annual Sysposium, Mary Kay O'Connor Process Safety Center, College Station, Texas, October 2011

[10] Donohue, S.K. and Dugan, J.B.; "Modeling the "Good enough to release" decision usin V&V preference structure and Bayesian belief networks", Reliability and Maintainability Symposium, 2003 Annual, 2003

[11] McNaught, K.R. and Zagorecki, A. "Using Dynamic Bayesian networks for prognostic modeling to inform maintenance decision making", IEEM 2009, Hond Kong, January 2010

[12] Dean, T. and Kanazawa, K., "A model for reasoning about persistence and causation", Computational Intelligence 1989; 5:142-150

[13] Bobbio A, Portinale L, Minichino M and Ciancamerla E., "Improving the analysis of dependable systems by mapping fault trees into Bayesian networks", Reliability Engineering and System Safety 71(3): 249-260

[14] Bearfield G and Marsh W, "Generalising event trees using Bayesian networks with a case study of train derailment", Available online at http://www.agenarisk.com/resources/white_papers/Generalising_Event_Trees.pdf , last modified: 23 February 2010, accessed: 7 January 2012

[15] Decision Systems Laboratory, University of Pittsburgh, GeNIe (Graphical Network Interface and SMILE (Structural Modeling, Inference, and Learning Enginer), Version 2.0 software, 2010. Available online at http://genie.sis.pitt.edu

(137b) Comparing the Performance of Compact Heat Exchange Reactor to CSTR for Fischer Tropsch Synthesis

Hani Gadalla[1], Zhijun Jia[1], Steven J. Vallee[1], Burtron H. Davis[2] and Dennis Sparks[2], (1)Chart Energy & Chemicals, La Crosse, WI, (2)Center for Applied Energy Research, University of Kentucky, Lexington, KY

A commercial Fischer Tropsch catalyst was studied in two Compact Heat Exchange Reactors (CHER) with different channel sizes, and in a continuous stir tank reactor (CSTR). The CHERs were constructed by Chart Energy and Chemicals with one reactor having a 5mm channels and the other having a 6mm channels, while maintaining the same catalyst volumeThe catalyst was crushed and sieved retaining particles ranging in size from 75-125μm. Testing was performed at the University of Kentucky Center for Applied Energy Research in Lexington, KY. The CO conversion, selectivity and hydrocarbon productivity were measured for each reactor over a wide range of operating conditions, as a function of temperature, pressure, H2:CO feed ratio, and flow rate. Both reactor types showed excellent temperature control, but the performances varied at different operating conditions. The larger channel CHER had a higher conversion than the 5mm channel, potentially due to slightly higher reactor bed temperature. The two CHERs approached and even exceeded the productivity of the CSTR for liquid hydrocarbons per gram of catalyst. At higher flow rates the methane selectivity increases at a lower rate for either of the CHERs than for the CSTR.

(124a) Inspection War Stories: Part 1

Gregory A. Cantley, Marathon Petroleum Company LP, Ashland, KY

The recent inspection of multiple distillation columns within Marathon Petroleum Company demonstrates the need for a thorough inspection of tower internals prior to startup to avoid failure. A good understanding of the process and objectives of the tower are required to ensure that column internals will perform as expected.

(124b) Is the Hydraulic Gradient on Sieve and Valve Trays Negligible?

Henry Kister, Process Engineering, Fluor Corp., Aliso viejo, CA

Hydraulic gradient estimates are important for troubleshooting channeling and multipass maldistribution problem on trays, yet there is no satisfactory method in the open literature to evaluate it. A common assumption in sieve and valve tray calculations is that the hydraulic gradient is negligible. This is supported by the literature leading method for hydraulic gradient calculations. In contrast, a rule of thumb we published in the 8th edition of Perry's Handbook, based on quantitative gamma scan data analysis, suggests that the hydraulic gradient on these trays may be significant when the liquid load is high and the flow path is long.

This paper looks further into the prediction of hydraulic gradients by comparing to published FRI data.

(124c) Troubleshooting and Improving the Operation of a Main Fractionating Tower

Lowell Pless, Tracerco, Pasadena, TX and Shane LaCasse, CHS Refining, Laurel, MT

Following a refinery turnaround, the Main Fractionator in the Ultra-Low Sulfur Diesel unit was not able to operate at expected rates. The Main Fractionator was field tested using radioisotope technology and studied for not being able to operate as expected. Gamma scans identified two sources of flooding originating from a pair of draw downcomers. Neutron backscatter technique was used to further investigate and confirm downcomer hydraulics to identify the root-cause of flooding in the Fractionator. The downcomer draw design was studied and determined to be conducive for plugging and subsequent flooding. A simpler downcomer design was proposed and implemented renovating an existing complicated piece of hardware. In addition the experience resulted in revamped turnaround procedures to avoid a repeat episode in the future.

(124d) Troubleshooting a Benzene Splitter Capacity Limitation

Waldo E. De Villiers, Shell Global Solutions (US) Inc., Houston, TX, Joanne Chew, Shell Global Solutions, Malaysia and Daisuke Honbou, Seibu Oil Co., Ltd.

A Benzene/Toluene Splitter in the aromatics Solvent Extraction Unit of a Refinery was experiencing a capacity limitation. The maximum achievable throughput corresponded to tray loads that fell short of the expected flood point. Attempts to increase throughput would result in increased pressure drop in the top part of the column together with higher than desired toluene in the benzene side-draw. This presentation will review the test runs, diagnostic scans and other troubleshooting steps that were performed to better understand the column constraints. One of the key observations was an unexpected high liquid level on the benzene partial draw tray. Careful consideration of the test run and other diagnostic data led to a number of recommended tray modifications. These modifications were implemented and following start-up the column was able to achieve significantly higher throughput without increased toluene in the benzene. One of the interesting aspects of this case study was the interaction that was observed between the column and side-draw receiver due to the location of an equalizing line nozzle.

Unique Ultrasonic Measurement Solutions

Anand N. Vennavelli
Fractionation Research, Inc.
vennavelli@fri.org

Jon Cloy
Marketing, GE Sensing
Jon.Cloy@ge.com

Paper # 124e, 2012 AIChE Spring Meeting, Houston, TX. *

INTRODUCTION

Ultrasonic clamp-on technology could provide valuable troubleshooting information and is particularly attractive because of the non-invasive nature of the technology.

In September 2011, GE's ultrasonic clamp-on liquid flow meters were installed on the Reflux and Bottoms pipes at the Fractionation Research, Inc. (FRI) unit with the following goals:

- Evaluate non-invasive clamp-on ultrasonic flow technology.
- Test novel applications of ultrasonic technology such as correlating sound speed with liquid density or composition.

So far, FRI completed two tests with the GE liquid clamp-on meters — one each in September 2011 and November 2011. The C_6/C_7 system at 23.5 psia was used in both tests.

In this paper, the preliminary results of the two tests are discussed. The flow measurements from the GE clamp-on meters are compared to FRI's measurements of the liquid flow. Sound speed, also measured from the GE meter, is correlated with liquid density and composition.

DATA COLLECTION

In September 2011, two clamp-on ultrasonic flow meters were installed on the bottoms and the reflux pipes of the FRI unit. Figures 1 and 2 show the photographs of the clamp-on flow meters. A transmitter-indicator was set up from which the flow and the sound speed values could be read.

Flow and sound speed readings were synchronized with liquid sample collections such that all flow and sound speed measurements were taken at steady-state and corresponded to the composition and process data collected for FRI's tests. Mass flow rates measured by FRI's Coriolis flow meters were converted to volumetric flow rates for comparison.

RESULTS AND DISCUSSION

The range of the reflux and the bottoms flow rates for the two tests are shown in Table 1. The November 2011 test had a narrower range than the September 2011 test. The flow rate ranges for both tests were determined by the ongoing mass transfer program at FRI.

Table 1: Flow rate ranges for the 2011 GE Liquid Clamp-on Tests. The ranges were determined by the ongoing mass transfer program.

Clamp-on Test	Bottoms Flow GPM	Reflux Flow GPM
Sep 2011	175–400	9–60
Nov 2011	145–228	31–55

Flow Measurement Results

The GE ultrasonic liquid flow meter measures the flow rate by measuring the sound transit time through the liquid. The transit time is based on the difference in the time taken by sound to travel up-stream and down-stream. The volumetric flow rate can be read directly from the transmitter-indicator.

The volumetric flow rates measured by GE's flow meter and FRI's measurement for the reflux flow are compared in Figure 3 in a parity plot. The plot shows an excellent agreement between the liquid flow rates measured by the ultrasonic flow meter and FRI's Coriolis meter. The mean absolute relative error between the reflux flow ultrasonic and FRI measurements was 5% for the September test data 6% for the November test data.

Similarly, the volumetric flow rates measured by GE's flow meter and FRI's measurement for the bottoms liquid flow are compared in Figure 4 in a parity plot. The mean absolute relative error between the bottoms flow ultrasonic and FRI measurements was 13% for the September test data and 14% for the November test data. The scatter observed in the bottoms flow measurements was greater than that in the reflux flow measurements, but still within the initial expectation of 15%.

Sound Speed Measurement Results

The liquid ultrasonic flow meter indicates both the flow rate and the sound speed through the liquid. Throughout the test, when reflux and bottoms were sampled, the sound speed through the liquid was also recorded.

The volumetric flow rate is determined from the sound transit time characteristics. However, the sound speed value by itself could provide information about the process liquid. One of the R&D goals of this work is to explore a correlation between sound speed and liquid density and sound speed and composition. Sound speed through a liquid mixture is affected by both temperature and composition. After accounting for the effect of temperature changes, the variation in sound speed can be directly attributed to composition changes.

The reflux and the bottoms temperature ranges are shown in Table 2. For most pure liquids, sound speed is inversely proportional to temperature. A temperature correction to sound speed was implemented in this report assuming that the inverse relationship between sound speed and temperature is also valid for the

Table 2: Temperature ranges for the 2011 GE Liquid Clamp-on Tests. The ranges were determined by the ongoing mass transfer program.

Clamp-on Test	Bottoms Temperature °F	Reflux Temperature °F
Sep 2011	229–238	149–171
Nov 2011	229–234	135–154

C_6/C_7 binary mixture. The reflux sound speed was calculated using Equation (1) at a basis of 165°F and the bottoms sound speed using Equation (2) at a basis of 230°F. Note that the choice of the basis is arbitrary.

$$\text{Reflux sound speed (ft/s)} = \frac{\text{Recorded Sound Speed (ft/s)} \times \text{Reflux Temperature (°R)}}{625\ \text{(°R)}} \tag{1}$$

$$\text{Bottoms sound speed (ft/s)} = \frac{\text{Recorded Sound Speed (ft/s)} \times \text{Bottoms Temperature (°R)}}{690\ \text{(°R)}} \tag{2}$$

The positive correlation between the sound speed and liquid density of the reflux and the bottoms, first observed in the September test, was confirmed in the November test as shown in Figures 5 and 6.

Similarly, the positive correlation between sound speed and liquid C_6 composition of the reflux and the bottoms, first observed in the September test, was also confirmed as shown in Figures 7 and 8.

In summary, the sound speed measurements correlate well with liquid density and composition changes. The GE liquid clamp-on transmitter can potentially be used for the C_6/C_7 system to detect changes in liquid density or as a clamp-on on-stream analyzer!

CONCLUSIONS AND FUTURE WORK

The ultrasonic flow meters showed potential to provide liquid flow measurements with 5–15% accuracy using a non-invasive clamp-on technology.

According to GE, more accurate flow measurements (up to 2% and up to 0.25% in some cases) may be obtained by using flanged-in invasive wetted sensors. However, the goal of this work was to study the effectiveness of clamp-on technology.

Considering that the flow measurements have been obtained using a non-invasive clamp-on device, the technology is industrially very valuable, especially for troubleshooting. Furthermore, since the flow meter has no moving parts, little to no maintenance or calibration is required.

The results of the September and November tests together showed that the ultrasonic measurements are repeatable and consistent. The November test results support all conclusions from the September test. The ultrasonic technology showed potential for detecting changes in liquid density and composition and therefore potential for use as an on-stream analyzer for the C_6/C_7 system.

Additional experiments and testing of novel applications of ultrasonic technology are planned.

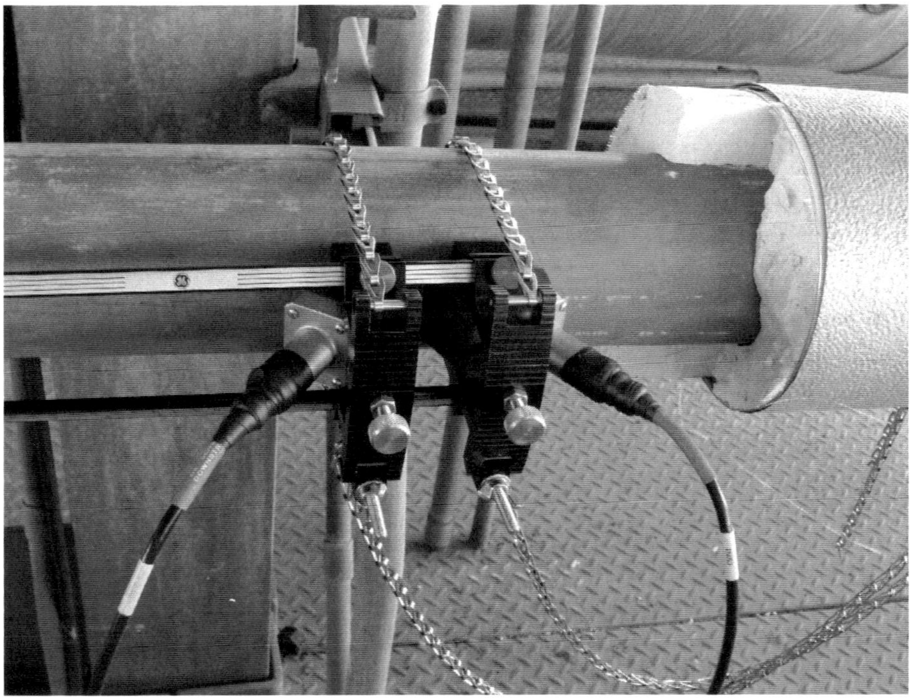

Figure 1: Photograph of the GE clamp-on ultrasonic flow meter on the reflux pipe.

Figure 2: Photograph of the GE clamp-on ultrasonic flow meter on the bottoms pipe.

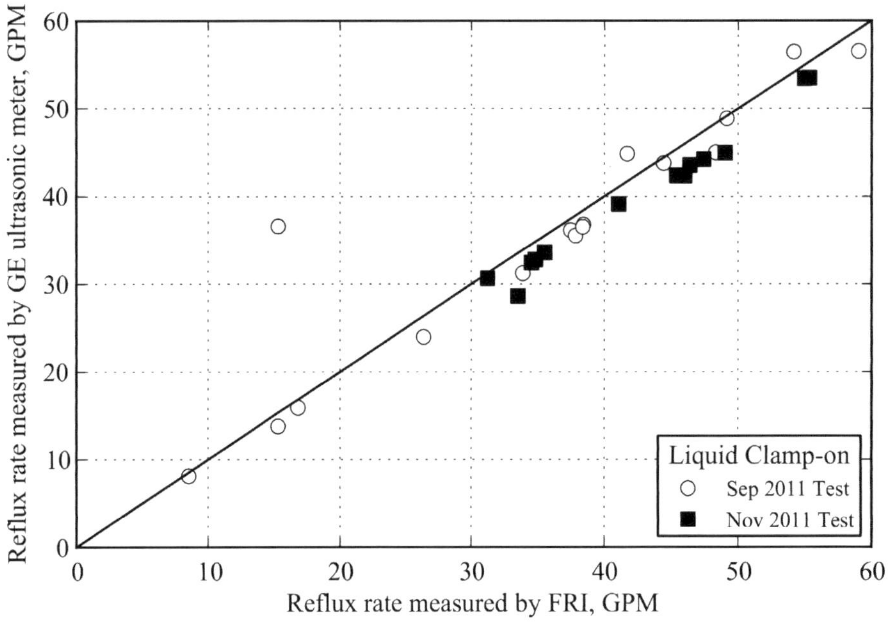

Figure 3: Parity plot of the reflux flow rate measurements

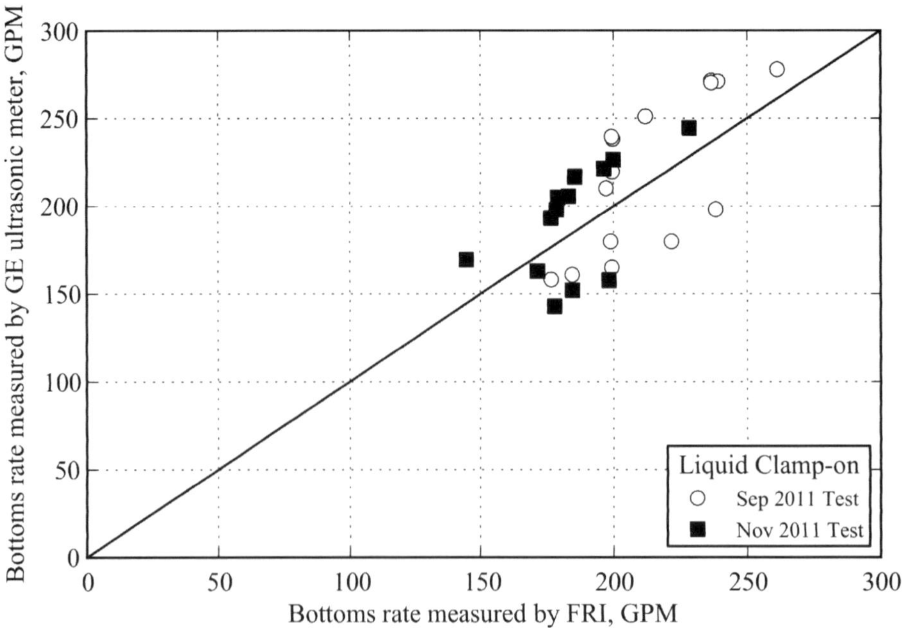

Figure 4: Parity plot of the bottoms flow rate measurements

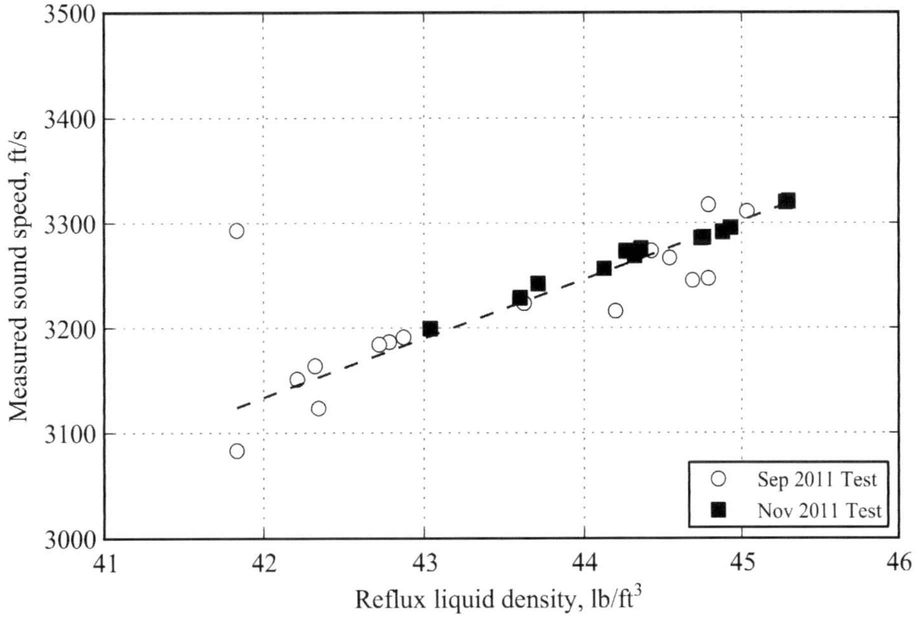

Figure 5: A positive linear correlation between the reflux sound speed and liquid density was observed. C_6/C_7 system at 23.5 psia. Reflux flow rates ranged from 9–60 gpm for the Septemer 2011 test and 31–55 gpm for the November 2011 test.

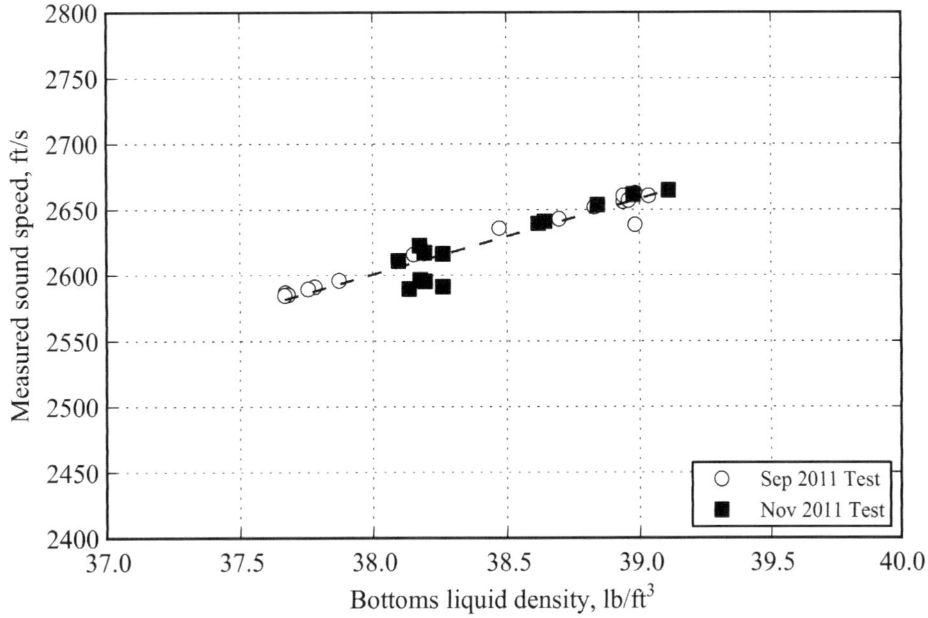

Figure 6: A positive linear correlation between the bottoms sound speed and liquid density was observed. C_6/C_7 system at 23.5 psia. Bottoms flow rates ranged from 175–400 gpm for the Septemer 2011 test and 145–228 gpm for the November 2011 test.

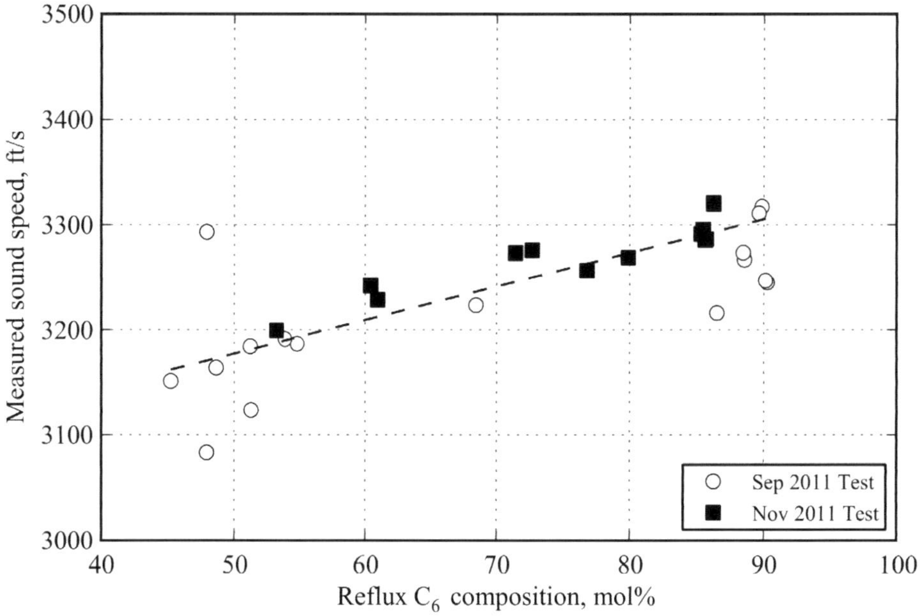

Figure 7: A positive linear correlation between the reflux sound speed and C_6 composition was observed. C_6/C_7 system at 23.5 psia. Reflux flow rates ranged from 9–60 gpm for the Septemer 2011 test and 31–55 gpm for the November 2011 test.

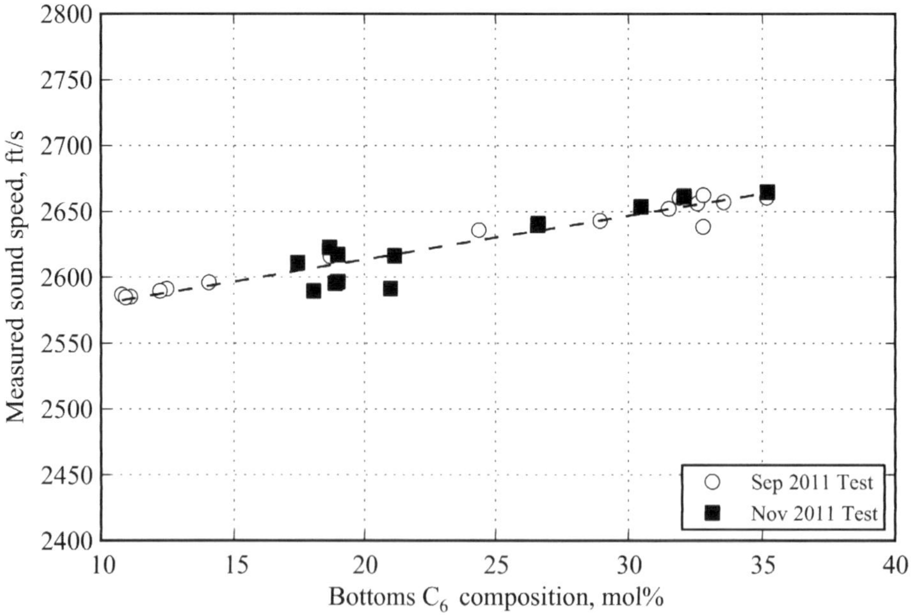

Figure 8: A positive linear correlation between the bottoms sound speed and C_6 composition was observed. C_6/C_7 system at 23.5 psia. Bottoms flow rates ranged from 175–400 gpm for the Septemer 2011 test and 145–228 gpm for the November 2011 test.

(124f) The Influence of Low Surface Tension on the Performence of Mist Eliminators

Wolfgang Heikamp, Rhodius GmbH, Weienburg, Germany, Michael Schultes, Raschig Ring Department, RASCHIG GMBH, Ludwigshafen, Germany and Attilio J. Praderio, LNG Technology Development, ConocoPhillips, Houston, TN

The influence of surface tension on the performance of mist separator

Wolfgang Heikamp, Rhodius GmbH

Attilio Praderio, Conoco Philipps

Prof. Dr. Michael Schultes, Raschig GmbH

In many chemical, refinery and petrochemical processes wire mesh mist separators are installed to remove fine droplets from gas streams. The design of these separators are mostly based on the following equation using a wire mesh characterizing K- factor,

$$Vc= K\sqrt{\frac{\Delta\rho}{\rho}} \quad \rho = \rho_L$$

Vc= design flow velocity and is a function of the L/V liquid/gas mass flow ratio, the density difference between liquid and gas $\Delta\rho$ and the liquid density ρ_L.

This equation (Souders - Brown) became an industrial standard to calculate a knitted wire mesh separators. There exist many proposals for the K factor in different applications.

Another design criteria was initially recommended by Brkholz, based on considerable test data, which can also be used for the estimation of the flooding velocity.

It has to be noticed that this equation does not take into account any surface tension effects and has been demonstrated only for air/water systems. Based on other gas/liquid separation processes it is known that low surface tensions (especially below 20 mN/m) reduce the flooding velocity.

In recent experimental studies with commercially available mist eliminators the influence of surface tension on the flooding velocity was investigated. A test rig was developed to study the flooding performance of mist eliminators with surface tensions from 68 mN/m down to 15 mN/m.

The results of these experimental investigations will be presented and compared to literature data. In addition an overview of future work will be given

(142a) Assessment of Minimum Energy and Operation Flexibility of Complex Dividing Wall Columns by Use of the Vmin-Diagram

Ivar J. Halvorsen, ICT, SINTEF, Trondheim, Norway, Sigurd Skogestad, Norwegian University of Science and Technology (NTNU), Trondheim, Norway, Igor Dejanovic, Chemical Engineering, University of Zagreb, Zagreb, Croatia and Zarko Olujic, Process & Energy Laboratory, Delft University of Technology, Delft, Netherlands

The dividing wall distillation columns (DWC) find increased use in industrial practice. The benefits are the combined savings in both energy and capital cost. Most of the industrial applications are three-product columns, but also 4-product DWCs have been constructed. A simulation study has shown that a four-product DWC with multiple partition walls for an aromatics feed can save up to 50% energy compared to conventional sequences. The internal arrangement is more complex and it is important to predict the necessary internal flow rates both for making proper sizing of the internals and also for control structure design. Thus, there is still a need to focus on development of simple-to-use methods for making assessment on how complex columns like the DWC will perform for any given application. This paper will demonstrate the usability of the Vmin-diagram method for quick assessment of column configurations for multicomponent separation.

The Vmin-diagram is obtained from feed data only, and gives a direct picture of the minimum vapor requirement for any specification of product splits. It is particularly well suited for fully thermally coupled distillation configurations like the generalized Petlyuk arrangements and dividing wall columns. The Vmin-diagram was first presented by Halvorsen and Skogestad (2003). The deduction was based on the simplifying assumptions of constant molar flows, constant pressure and constant relative volatility. It can also be constructed from simulations on a real mixture.

The minimum vapor flow rate in a DWC is given by the most difficult of the split between two adjacent product groups as if each split were to be carried out in a standard two-product column. For sharp split specifications, these are found as peaks in the Vmin-diagram. In addition, a base solution for the internal flow rates in each internal column section can be found directly.

In a DWC we fortunately have some tolerance in how to set the internal flows since there is normally more than a single optimal operating point. This flexibility arises when there are different minimum required flow rates in the upper and lower part of the arrangement. This tolerance can be explored in the Vmin-diagram and be used to specify operational margins on the internal flow rates. Knowledge of the operational tolerances for optimal operation is important in order to design and operate a column that obtains the theoretical energy savings in practice.

(142b) Experimental Evaluation of a Concentric Tray HIDiC Performance

Zarko Olujic, Sander Tromp, Joost Laven, Coen Dusseljee and Andrzej Stankiewicz, Process & Energy Laboratory, Delft University of Technology, Delft, Netherlands

An internally heat integrated distillation column (HIDiC) is essentially a heat pump assisted distillation column that in conjunction with diabatic operation, i.e. a uniform distribution of heat transfer duty along the integrated parts of a hot rectification and a cold stripping section, enables minimization of energy requirement in demanding separations. It is a proven, but regarding the hardware a highly demanding energy conservation concept. Indeed, uncertainties related to both design & construction and operation of a HIDiC are the main reason for reluctance on the side of potential users that diminishes prospects for practical implementation.

In present, technical feasibility study a concentric column containing an 0.3 m diameter rectification section placed inside an 0.8 diameter stripping section, with five fully thermally integrated sieve trays, has been used to simulate experimentally operation of a HIDiC. Two heat integrated sections have been operated as two individual columns in parallel at total reflux conditions, using appropriate mixtures of cyclohexane and n-heptane. The inner, rectification column (section) of a HIDiC was operated at higher pressures, while the outer, stripping section was operated at atmospheric pressure. The heat transfer from hot rectification column (section) to cold stripping column (section) occurred via the surface of the inner column wall and the heat panels placed above active (bubbling) area of the annular sieve trays in the stripping column. The photograph shows design of equipment.

The performance of the small diameter inner column (cartridge sieve trays) appeared to be more sensitive to the rate of internal heat transfer than the outer column, exhibiting a deteriorating trend in efficiency with increasing column vapor load. The outer column equipped with annular sieve trays exhibited a stable performance over the whole operating range and as expected the presence of heat transfer panels appeared to be beneficial to both hydraulics and mass transfer efficiency. However, the overall heat transfer coefficient exhibits a progressive increase with increasing vapor load, indicating that at lower loads insufficient wetting, i.e. vaporization occurs, which is a concern. This and other concentric tray HIDiC design and operation relevant considerations are discussed.

(142c) Kettle Reboiler Windows

T.J. Cai, Simon Chambers, A. Y. Ogundeji and A.N. Vennavelli, Fractionation Research, Inc., Stillwater, OK

Fractionation Research, Inc. (FRI) recently installed two windows on an industrial-size kettle reboiler. Using those windows, data were collected regarding pool depths, entrainment, flow patterns and nucleation sites. Several binary systems were employed. A wide range of steam pressures were employed. Video footage will be shown.

(142d) An Innovative Energy Saving Based on Self-Heat Recuperation for Distillation Process

Yasuki Kansha, Akira Kishimoto and Atsushi Tsutsumi, Institute of Industrial Science, The University of Tokyo, Tokyo, Japan

Although distillation columns have been widely used in separation processes based on vapor/liquid equilibria in petroleum refineries and chemical plants, the distillation process consumes a massive amount of energy required for the latent heat of the phase change, resulting in the emission of a large amount of CO2.

To overcome this problem, an innovative energy saving based on self-heat recuperation for distillation processes is proposed in this research; the energy saving occurs via internal heat circulation of the process. It consists of a heat circulation module and a distillation module in which the heating and cooling loads are balanced by energy/exergy analysis. The self-heat recuperation technology proposed in our previous studies is adopted, and thus, the heat of condensation and the cooling heat are recuperated by compressors, and exchanged with the heat of vaporization and the heating heat by employing heat exchangers in both the heat circulation and distillation modules.

The self-heat recuperation technology helps to achieve a drastic energy saving for distillation of more than 80% in comparison with a benchmark distillation involving the use of an external heat source and a reboiler for heating.

APPLICATION OF RESPONSE SURFACE METHODOLOGY TO OPTIMIZE THE RETROFIT OF A SIDE COLUMN TO A DWC

Nguyen Van Duc Long and Moonyong Lee, Yeungnam University, Gyeongsan, South Korea

Introduction

Distillation plays an important role in the chemical process industries and consumes the largest amount of energy with an estimated 3% of the world's energy consumption [1]. Motivated by this large energy requirement for distillation, researchers have developed various column arrangements that can bring in savings in both energy and capital cost [2]. Any reduction of energy consumption will not only bring economical benefits but also environmental benefits in terms of reduced usage of fossil fuels and its associated emissions.

A side distillation column can be used to separate any nonazeotropic multicomponent mixture into three products [3]. Therefore, it can often substitute for a sequence of two columns. However, the purity of the side distillation is restricted by thermodynamics and by the nature of the distillation process. These columns are usually appropriate either as prefractionators (the side stream is fed to another column for further separation) or to generate recycle streams when there is no strict requirement of the recycle composition. They are also broadly used in the petroleum industry, where blends are produced rather than pure products.

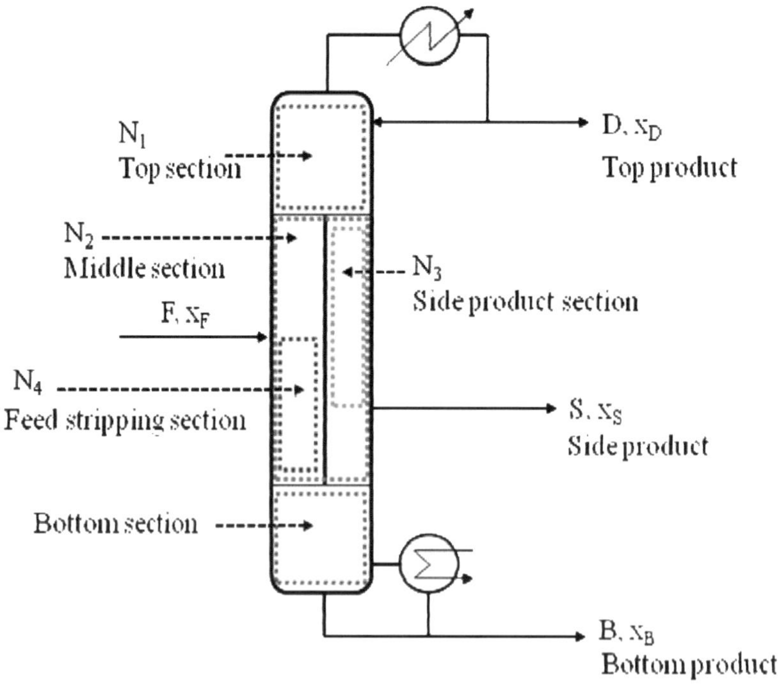

Figure 1. A dividing wall column.

There is a significant amount of literature analyzing the relative advantages of dividing wall column (DWC) options, schematically drawn in Fig. 1, offering a great chance at reduced energy consumption. These studies have shown that DWC systems are capable of achieving energy savings of up to 30% over conventional direct and indirect distillation sequences [4-8]. Besides, the investment and space saving can be gained because the number of reboilers, condensers and associated equipments such as pumps, their supports [2]. Therefore, DWC is attractive to many chemical and related industries in the current scenario of competition and environmental concerns, in order to reduce energy usage for distillation.

Retrofit a side distillation column to a DWC has potential for reducing energy. The successful revamp of a conventional tray Xylenes Distillation Column was accomplished by ExxonMobil [9]. The column takes a xylene sidestream from reformate to feed an aromatics plant to make higher value products. Changing the column configuration to a DWC resulted in improved product purity and significant energy savings. The project cycle is detailed including process design, hydraulic design, mechanical design, manufacture, testing and site work. An investment with payback period of less than 3 years is favorably considered by the management.

Recently, Premkumar & Rangaiah (2009) and Long (2010) studied the potential of retrofit conventional 2-column systems and side for separating ternary mixtures into three products to DWCs [2,7]. DWCs were optimized only one variable at a time, keeping constant the remaining ones. For each chosen variable value, the internal vapor and liquid flow to the prefractionator were varied to optimize energy consumption. The interactions between variables could not be identified and quantified by such a technique. Latterly, Long and Lee proposed the response surface methodology (RSM) and factorial design for DWC structure design and optimization, which is easy and efficient to implement using Hysys and Minitab [10-12]. These methods require a little computational effort while it allows control variables to be optimized simultaneously. Both of them develop the mathematical models that permit assessment of the relevance and statistical significance of variables being studied [13]. They facilitate the evaluation of interaction effects between variables. The aim of the present work is to show potentiality of retrofit side distillation column to a DWC. The response surface methodology was employed to minimize the reboiler duty.

Design and Optimization Methodology

Design

The key to a successful retrofit lies in maximizing utilization of the existing equipment, while simultaneously minimizing the new hardware so as to abbreviate capital costs [14]. The side column is now modified by adding a wall in the middle section while maintaining the same number of trays. Initial DWC structure design involved applying a shortcut design procedure based on a conventional column configuration [7,12]. In this conventional configuration, the first column corresponds to the prefractionator section in the DWC. The rectifying section of the second column and the stripping section of the third represent the top and bottom sections of the DWC, respectively. The stripping section of the second column and the rectifying section of the third are equivalent to the dividing wall section of the DWC. The bottom stream from the second column and the top stream from the third column correspond to the side stream of the DWC. Consequently, the DWC structure can be divided into four sections: the prefractionator section for feed mixture, the top and bottom sections above and below the dividing wall section, and the dividing wall section.

Response surface methodology

Response surface methodology (RSM) is also employed to optimize the retrofit the side column to dividing wall column. After determining the preliminary ranges of the variables through single-factor testing, a Box-Behnken design was employed under the response surface methodology to analyze how the variables interacted and optimize the system in terms of reboiler consumption and cost saving. Simulation run data were fitted to a second-order polynomial model and regression coefficients were obtained. The generalized second-order polynomial model used in the response surface analysis is as follows:

$$Y = \beta_0 + \sum_{i=1}^{k} \beta_i X_i + \sum_{i=1}^{k} \beta_{ii} X_i^2 + \sum\sum_{i<j} \beta_{ij} X_i X_j + \varepsilon \qquad (1)$$

where Y is the predicted response, X_i are the uncoded or coded values of the variables, β_0 is a constant, β_i, β_{ii} and β_{ij} are the coefficients of the linear, quadratic and interactive terms, respectively, and ε is the error term.

Case Study

Existing Side Distillation Column

Figure 2 illustrates the existing distillation column and current operating conditions. The existing side column is valve-trayed type of 3.8 m diameters, with 65 trays [15]. The feed composition, temperature, and pressure are described in Table 1. The simulation work was performed using simulator Aspen HYSYS V7.1. The NRTL-HOC property method that uses the Hayden-O'Connell equation of state as the vapor phase model and NRTL for the liquid phase was used for the prediction of the vapor-liquid equilibrium (VLE) of these simulations. Dimerization affects VLE, vapor phase properties, such as enthalpy and density, and liquid phase properties, such as enthalpy. The Hayden-O'Connell equation reliably predicts the solvation of polar compounds and dimerization in the vapor phase that occurs with mixtures containing carboxylic acids [16]. From the base case simulation model, it shows that the energy consumption of this column is 22.50 MW.

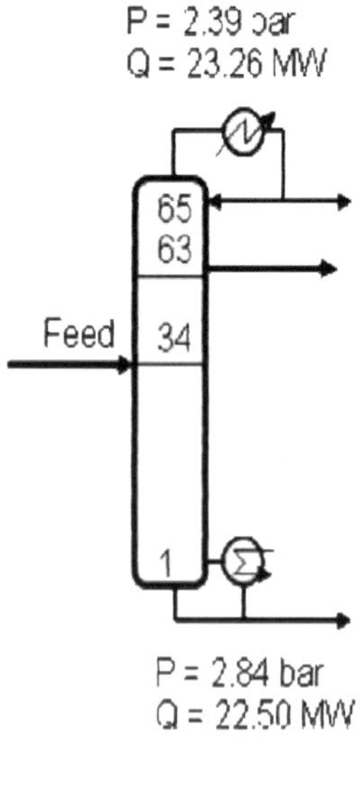

Figure 2. Simplified flow sheet illustrating the existing side distillation column.

Column hydraulics, as well as flooding, has to be considered. When the downcomer is small or the downcomer backup is high, downcomer flooding occurs, causing liquid accumulation in the tray above, while the rising of the vapor velocity causes entrainment. To determine the max flooding of a particular column, the rating mode is simulated with column internal specifications such as type of trays, column diameter, tray spacing, and number of passes. The side column was designed with a load which is at near 85% of the load at the flooding point to prevent flooding in the columns.

Table 1. Feed mixture condition and product requirement

Feed Conditions	
Component	Mass Flow (kg/h)
Formic Acid (FA)	36.29
Water	15.88
Acetic Acid (AA)	51848.41
Propionic Acid (PA)	313.43
Temperature (°C)	173.8
Pressure (bar)	4.46
Product Requirement	
Formic Acid (FA)	<1000 ppm
Water	<1000 ppm
Propionic Acid (PA)	<300 ppm

Dividing wall column

Table 2. Factors' coded levels for RSM

Factor	Levels		
	-1	0	1
top section (N1)	1	5	9
middle section (N2)	36	43	50
Side product location (N3)	1	3	5
Feed stripping location (N4)	1	5	9

In this paper, the retrofit project purpose is to reduce energy requirement, with the following assumptions:

a. The purity and recovery of AA is kept constant.
b. Specification of product: - FA and water concentration must be less than 1000 ppm
   - PA concentration is less than 300 ppm
c. Steam @ 150 psig (11.36 bar): 880 USD/KW.year [17]
d. Remained year: 8 years
e. Column material: stainless 316
f. New trays are installed in the dividing wall section

Table 2 shows the factors and levels used in this case study. 27 simulations were run to optimize 4 parameters of the DWC structure. For each run, internal vapor (F_V) and liquid (F_L) flows to the prefractionator were varied to minimize the reboiler duty. The minimum reboiler duty was found at coded levels of the number of trays in the top, middle, side product, and feed stripping section of -1.000, -0.3737, -1.000 and -1.000, respectively (Fig. 3).

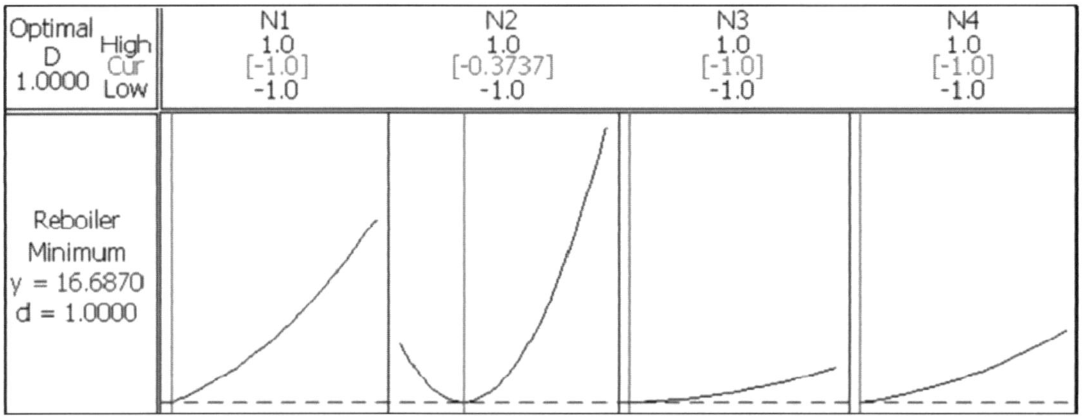

Figure 3. Optimization plot by the RSM.

Under these conditions, the minimum reboiler duty was predicted as 16.69 MW. The natural values of the valuables can be derived from the coded levels. F_V and F_L then were optimized to maximize the total annual cost saving, while maintaining sufficient product purity. Figure 4 shows a simplified flow sheet illustrating the resulting DWC system. The result from rigorous simulation also revealed that the DWC can save up to 25.16% of energy consumption, which showed good agreement with the predicted result by the RSM. The predicted maximum costs saving by the DWC is 3.89 million USD after 8 years. Simple payback period can be calculated as additional capital costs divided by saving per year. In this case the value is 2.2 months due to high energy cost. The chemical/petrochemical plants usually have an annual maintenance period of 10 – 14 days, during which major repair and replacement are carried out. The retrofitting discussed above has been assumed to take place within this maintenance period, thus production loss during the retrofitting time is not considered [2]. Note that the temperatures on the bottom section of the side column and DWC were almost same (165°C), medium pressure steam are used in both columns. Max flooding reduces from 87% to 67%, i.e. retrofit of side distillation column gives some chances for increasing the plant production capacity.

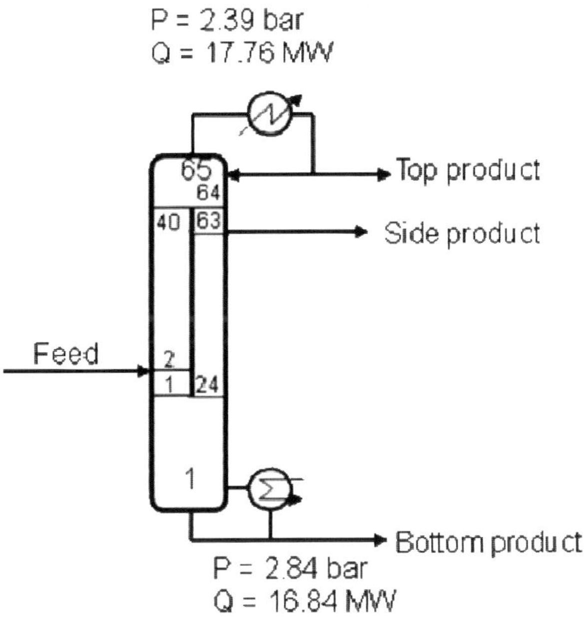

Figure 4. Simplified flow sheet of the resulting DWC system using RSM.

Conclusions

This paper has focused upon the retrofit the side column to a DWC, which maximizes utilization of the existing equipments and operate the plant more efficiently, possibly with minor modifications and small investment. The shortcut method, utilizing the three conventional column configuration structurally equivalent to the DWC, was used to determine the proper initial structure for each DWC in a simple manner. The response surface methodology was employed for retrofit design and optimization. The optimal structure design of the DWC has been considered in terms of the internal recycle flow distribution around the dividing wall, the number of trays in the top, middle, side product and feed stripping sections. One case study was analyzed and the results showed that RSM is easy and efficient to employ for retrofit design and optimize. The retrofitting DWC structure was efficiently optimized with a little computational effort. The predicted results by the RSM showed a good agreement with the actual ones by rigorous simulation. The resulting DWC leads to significantly reduced energy costs over side distillation column.

Acknowledgments

This work also was supported by the Human Resources Development Program of Korea Institute of Energy Technology Evaluation and Planing (KETEP) grant (No. 20104010100580) funded by the Korean Ministry of Knowledge Economy.

References

1. G. Hewitt, J. Quarini, M. Morrell, More efficiency distillation, *Chem. Eng.* (1999).
2. R. Premkumar, G.P. Rangaiah, Retrofitting conventional column systems to dividing wall columns, *Chem. Eng. Res. Des.* 87 (2009) 47-60.
3. K. N. Glinos, M. F. Malone, Design of sidestream distillation columns, *Ind. Eng. Chem. Process Des. Dev.,* 24 (1985) 822-828.
4. Victoria E. Tamayo-Galván, Juan Gabriel Segovia-Hernández, Salvador Hernández, Julián Cabrera-Ruiz, J. Rafael Alcántara-Ávila, Controllability analysis of alternate schemes to complex column arrangements with thermal coupling for the separation of ternary mixtures, *Com. Chem. Eng.*, 32, 3057-3066, 2008.
5. Hern´andez, S., & Jim´enez, A. (1999a). Design of energy-efficient petlyuk systems. *Computers and Chemical Engineering*, 23(8), 1005.
6. Sotudeh, N., & Shahraki, B. H. (2008). Extension of a method for the design of divided wall columns. *Chemical Engineering Technology*, 31, 83.
7. N. V. D. Long, Lee, S. H. & Lee, M. Y. (2010). Design and optimization of a dividing wall column for debottlenecking of the acetic acid purification. *Chemical Engineering and Processing*, 49(8), 825-835.
8. C. Triantafyllou, R. Smith, The design and optimization of fully thermally coupled distillation, *Trans. Inst. Chem. Eng.* 70 (Part A) (1992) 118-132.
9. B. Slade, B. Stober, D. Simpson, Dividing wall column revamp optimizes mixed Xylenes production, *IChemE Symp.* Series 2006, 152.
10. N. V. D. Long and M. Y. Lee, Dividing wall column structure design using response surface methodology, *Computers & Chemical Engineering*, 37 (2012) 119-124.
11. N. V. D. Long and M. Y. Lee, Design and optimization of a dividing wall column by factorial design, *Korean Journal of Chemical Engineering*, In Press (2012).
12. N. V. D. Long and M. Y. Lee, Improvement of the Deethanizing and Depropanizing Fractionation Steps in NGL Recovery Process Using Dividing Wall Column, *Japan Journal of Chemical Engineering*, In Press (2012).
13. César Ricardo Teixeira Tarley, Gustavo Silveira, Walter Nei Lopes dos Santos, Geraldo Domingues Matos, Erik Galvão Paranhos da Silva, Marcos Almeida Bezerra, Manuel Miró, Sérgio Luis Costa Ferreira, Chemometric tools in electroanalytical chemistry: Methods for optimization based on factorial design and response surface methodology, *Microchemical Journal* 92 (2009) 58–67.
14. K.A. Amminudin, R. Smith, Design and optimization of fully thermally coupled distillation columns. Part 2: Application of dividing wall columns in retrofit, *Trans. IChemE.* 79 (2001) 716-724.
15. PEP Report, Acetic Acid by Low Pressure Carbonylation of Methanol, November, 1994.
16. Aspen Technology, Aspen Physical Property System, version number V7.1, January 2009.
17. Biegler, L.T., Grossmann, I.E. & Westerberg, A.W. (1997). Systematic methods of chemical process design. Prentice Hall Inc., New Jersey, 110-141.

A STUDY ON SEPARATION PROCESS IN FLNG

Nguyen Van Duc Long and Moonyong Lee, Yeungnam University, Gyeongsan, South Korea

Introduction

The primary use of natural gas (NG) is as a fuel; it can also be a source of hydrocarbons for petrochemical feed stocks and a major source of industrially important elemental sulfur [1]. Its clean burning and ability to meet stringent environmental requirements have raised the demand for natural gas [2], which is supporting the rapid growth of liquefied natural gas (LNG) production capacity. Much of the world's gas reserves are in offshore fields [3], though onshore LNG processing is generally favored [4-8]. Recently, offshore FLNG (floating LNG) and LNG-FPSO (floating, production, shipping and offloading) services are beginning to be explored instead of land-based LNG plants [9]. FLNG plants must be safe, compact and energy efficient.

Several analyses of dividing wall columns (DWCs, Figure 1) for ternary separations have shown that they can achieve energy and space savings of up to 30% over conventional direct and indirect distillations [10-12]. DWCs allow reversible splits with no part of the separation performed twice, the main reason for their superior energy efficiency [13]. However, their design is more complex than conventional arrangements because of the greater number of degrees of freedom [14] that interact with each other and need to be optimized simultaneously.

Optimization studies rely on statistical approaches, with response surface methodology (RSM) being routinely used in several biotechnological and industrial multivariable processes [15-16]. In RSM, the Box-Behnken design only has three levels (low, medium, and high, coded as -1, 0, +1) and requires a small number of experiments or simulation runs. It is more efficient and easier to arrange and interpret than other methods.

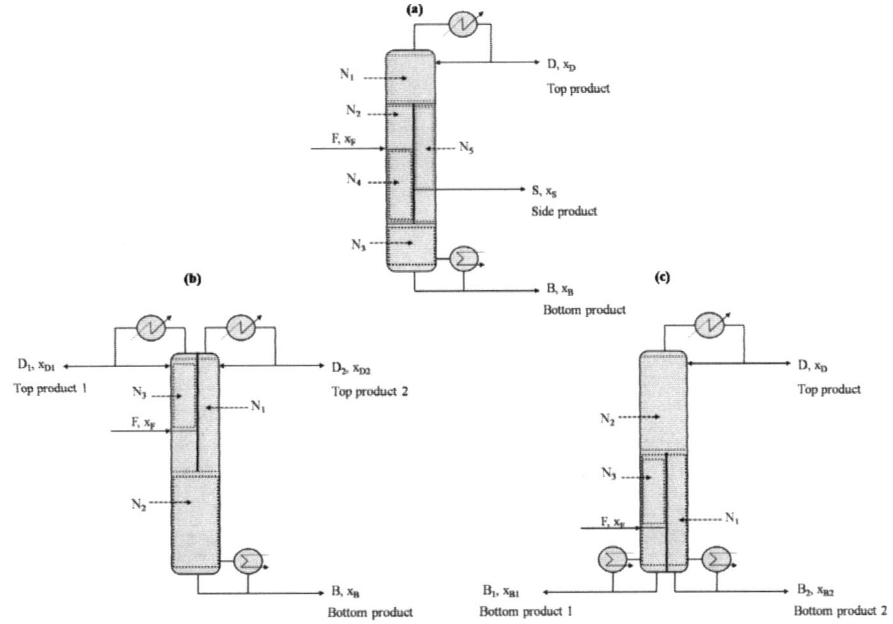

Figure 1. Dividing wall columns (DWCs): (a) conventional DWC; (b) top DWC; (c) bottom DWC.

This work proposes a new, simple, compact and power-efficient NGL recovery sequence for offshore FLNG. In the proposed NGL recovery process, feed splitting and a top dividing wall column were employed to maximize energy efficiency and plant compactness.

Analysis of Onshore LNG and Offshore FLNG Plants

A traditional onshore plant for offshore gas reserves usually includes an offshore platform for dehydration and compression, large pipelines to shore, an onshore liquefaction plant, and a harbor to accommodate purpose-built LNG carriers [17]. A floating LNG (FLNG) plant eliminates the platform, pipelines, and harbor, greatly reducing infrastructure requirements. The savings can be significant depending on the length of the pipeline and the cost of the harbor. Even for onshore gas fields, FLNG may also be less expensive than onshore facilities if extensive site preparation is required and there is no local infrastructure or natural harbor. It can also lessen a facility's environmental impact, particularly if the potential onshore plant would be located near populated areas. A FLNG plant can be moved to a new location after the depletion of the original gas field [18].

However, a FLNG facility's production capacity is limited by the deck space required for equipment, constraining throughput and economics, especially for projects on larger gas fields [19]. FLNG projects are more difficult to expand, as expansion is only achievable through the installation of additional ‘stand alone’ facilities. The expansion of a onshore plant would require less capital and materials, with only incremental additional environmental impact as additional common-use infrastructure such as power generation or export jetties may not be needed. A FLNG plant would have higher operating and maintenance costs than an onshore plant, with very limited operational synergies available for multiple FLNG facilities compared with an onshore plant operating multiple gas trains. FLNG plants are therefore most suited for developing small, remote gas fields that are difficult to develop commercially by conventional methods (often referred to as stranded gas assets).

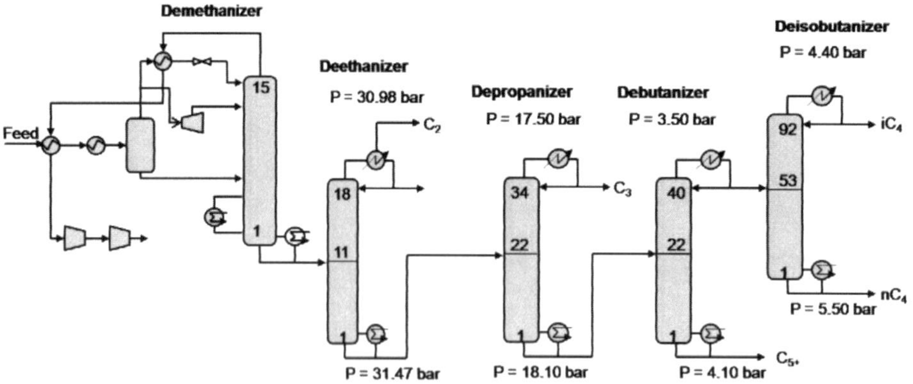

Figure 2. The separation train of typical onshore NGL processing.

Onshore liquefaction plants employ complex, high-efficiency liquefaction processes [20] that use mixed or cascade refrigerants with separate pre-cooling. However, offshore applications require simpler systems due to the challenges of moving processes offshore [18]. An offshore plant must be able to withstand waves, most notably the distillation and separation columns required for condensate stabilization, LPG extraction and acid gas removal in feed gas pretreatment. FLNG plants require higher levels of safety due to their compactness; interlocks, emergency shutdown, fire and gas detection, and fire fighting systems should be carefully considered from the conceptual stage [20].

Heat exchangers, cold-boxes and columns should be designed to withstand motion, making packed type column designs suitable [20]. This paper considers the potential process and the improvements that could be realized by separation. During onshore processing, liquid hydrocarbons recovered from NGL are typically separated into relatively pure ethane (C_2), propane (C_3), isobutane (iC_4), normal butane (nC_4), and gasoline (C_{5+}) products. This is conventionally done by the sequential distillation of C_2, C_3 and C_4 from gasoline, followed by further distillation of iC_4 from nC_4 (Figure 2) [4-8]. The similar boiling points of iso- and normal butane make their separation require many trays with a high reflux ratio. Due to restrictions of column height, a deisobutanizer would likely be excluded from an FLNG plant. In addition, the number of products and product tanks should also be minimized because of the limited space and safety concerns.

Proposed Base Sequence

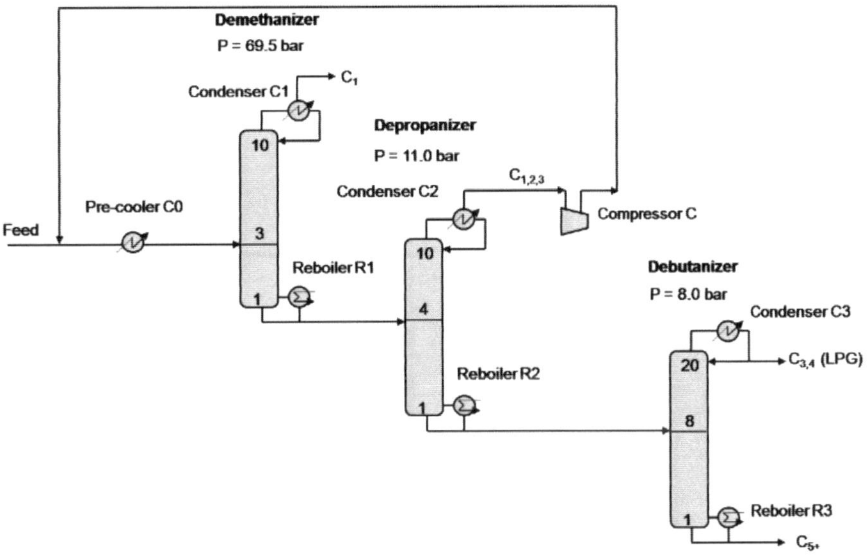

Figure 3. The separation train of three conventional columns.

A new NGL recovery process is proposed for FLNG facilities with the following constraints and assumptions:

a. The maximum allowable number of trays of each column is limited to 21 to limit the motion of wind and waves and to avoid excessive bending moments.
b. All columns are designed as packed type columns for greater stability against motion.
c. The operating velocity of all columns is near 80% of the flooding velocity.
d. The product specifications are as follows:
   - C_1: 89% (top of demethanizer)
   - LPG: 98% (top of debutanizer)
   - C_{5+}: 99% (bottom of debutanizer)

Based on the feed conditions and the product specifications, simulations were performed to quantify energy consumption. The demethanizer, with 10 theoretical trays, was designed for use at 69.5 bar (Figure 3). The sharp separation of methane from C_{2+} increases energy consumption. Therefore, a depropanizer column is used to recycle light components in the bottom stream of the demethanizer. The 20-tray debutanizer column was designed for use at 8.0 bar, as commercial LPG can be condensed with cooling water at this pressure.

The Peng-Robinson equation of state supporting the widest range of operating conditions and the greatest variety of systems was used to predict the simulations' vapor-liquid equilibria [21]. The Lee-Kesler method of enthalpy calculation was employed.

Improvement of Proposed Base Sequence

Feed Split

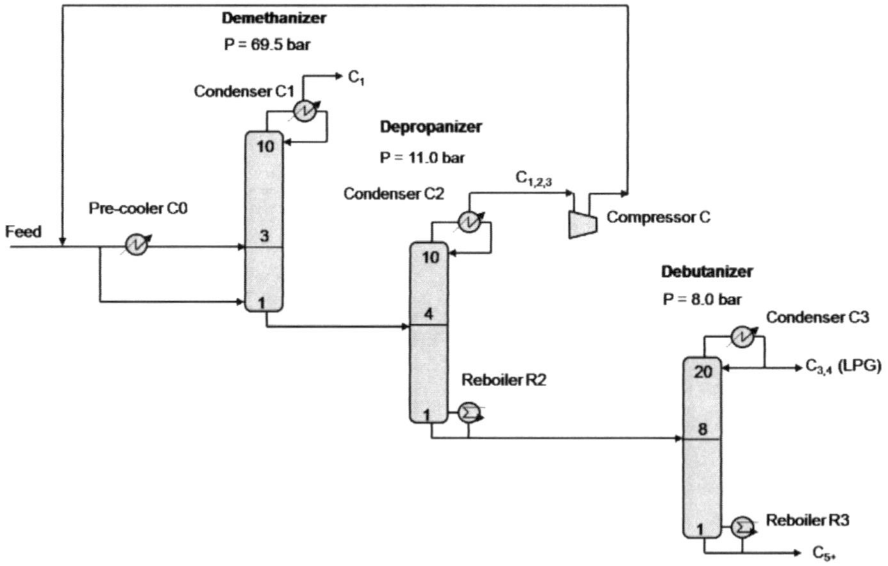

Figure 4. The separation train of three conventional columns with feed splitting.

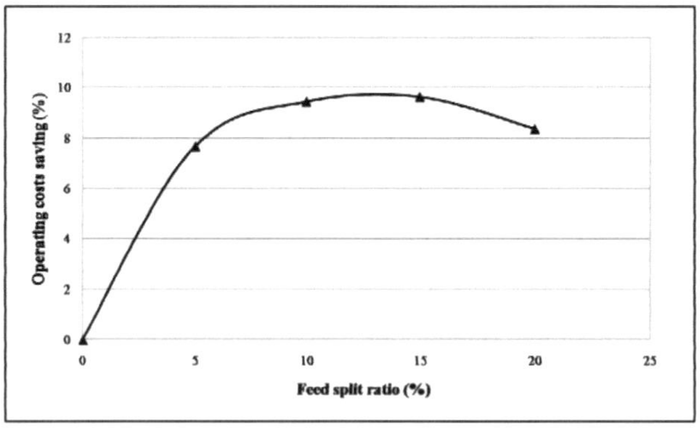

Figure 5. Effect of feed split ratio on the operating cost of the demethanizer and depropanizer.

Instead of pre-cooling the entire feed stream, it can first be divided into two streams: one is cooled before entering the column like a feed, and one that is fed into the column's bottom (Figure 4). This sequence can eliminate one reboiler in the demethanizer and save energy in pre-cooler C0. However, energy consumption is increased in condenser C1 in the demethanizer and the reboiler in the depropanizer. Note that the refrigeration used in precooler C0 and condenser C1 are at low temperatures. Therefore, adjustment of the feed split ratio is important in the optimization (Figure 5). A feed split ratio of 15% is optimal, reducing reboiler duty and operating costs by 47.43% and 9.64%, respectively.

Top Dividing Wall Column (TDWC)

In this case, two different cooling sources, water and refrigeration, are employed. Due to the high cost of refrigeration, integrating two columns into one dividing wall column with only one condenser is not efficient. Therefore, a top dividing wall column is considered here to integrate the depropanizer and the debutanizer for improved energy efficiency.

Optimization Methodology

The main design variables of internal vapor flow to the prefractionator (F_V), and the numbers of trays in the top (N1), botttom (N2), and feed rectifying (N3) sections were optimized. The objectives of the optimization were the minimization of the total number of trays (N) and the reboiler duty (Q). The optimization was constrained by the product purities and recoveries.

$$\min(Q,N) = f(N1,N2,N3,F_V) \tag{1}$$

Multiple response optimization can be undertaken by forming a constrained optimization problem with one of the responses treated as the objective of a constrained optimization problem and other responses treated as the constraints where the constraints' boundaries are determined by the process design engineers. Here, reboiler duty is considered the objective. The numbers of trays are the constraints and depend on the design engineers. Note that the number of trays in each column must be less than 21 trays.

A Box-Behnken design was employed under the response surface methodology to analyze how the variables interacted and to optimize the system in terms of reboiler duty and the number of trays. After determining the variables' preliminary ranges through single-factor testing, simulation run data were fitted to a second-order polynomial model and regression coefficients were obtained. The generalized second-order polynomial model used in the response surface analysis is as follows:

$$Y = \beta_0 + \sum_{i=1}^{k} \beta_i X_i + \sum_{i=1}^{k} \beta_{ii} X_i^2 + \sum\sum_{i<j} \beta_{ij} X_i X_j + \varepsilon \tag{2}$$

where Y is the predicted response, X_i are the uncoded or coded values of the variables, β_0 is a constant, β_i, β_{ii} and β_{ij} are the coefficients of the linear, quadratic and interactive terms, respectively, and ε is the error term. MINITAB software was used for response surfaces fitting.

TDWC Configuration

After initially setting the TDWC structure using a shortcut method [12,14], it was then optimized using response surface methodology. The DWC parameters were optimized over 15 simulation runs. For each run, the internal vapor flows to the prefractionator were varied to meet the required product purity and recovery. The optimum reboiler duty saving corresponds to the number of trays, increasing rapidly as the number of trays increases from 19 to 22. Therefore the maximum allowable 20 trays were employed in the design of the TDWC.

The greatest reboiler duty saving (11.89%) was predicted to have coded levels of -0.8192, 0.8192, and -1 corresponding to the numbers of trays in the top, bottom, and feed rectifying sections, respectively. The variables' natural values can be derived from the coded levels. The optimized TDWC system (Figure 6) had a simulated saving of 11.79%, in good agreement with the predicted value.

This column can be used to integrate two columns, whose condensers are cooled by different coolants; particularly the the depropanizer requires refrigeration, while the debutanizer can condense the top vapor stream under water cooling. The simulation results show that the TDWC can save up to

35.09% of refrigeration costs compared with a conventional column sequence. The duties of condenser (C3) and the reboiler can be reduced by 21.42% and 11.79%, respectively. Using a 1.5 m diameter TDWC could reduce energy consumption by 23.62% in terms of TAC compared with a conventional column sequence. Therefore, this kind of column can reduce energy consumption (refrigeration energy and reboiler and condenser duty) and investment costs (replacing two columns with one and eliminating a reboiler).

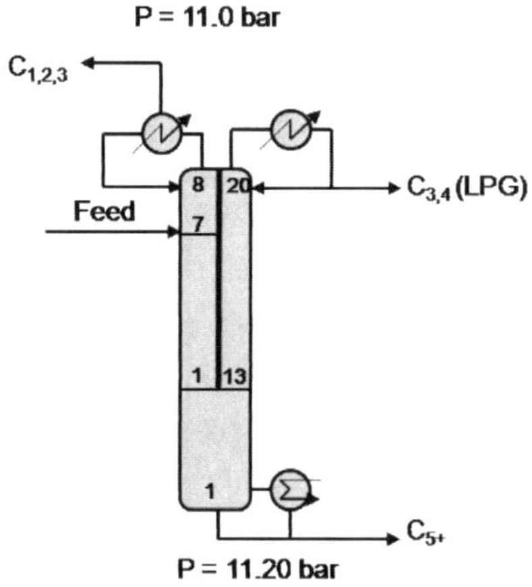

Figure 6. The TDWC system.

Conclusions

A new, energy-efficient and compact NGL recovery is proposed for offshore FLNG plants. Feed splitting could eliminate the demethanizer reboiler and reduce precooling energy. Operating costs were minimized at a feed split ratio of 15%. A TDWC allowed more energy-efficient depropanization and debutanization. A shortcut method determined a suitable initial structure of the TDWC, which was then optimized by response surface methodology. The proposed method could be simply and efficiently implemented in HYSYS and Minitab. Simulated results agreed well with predicted values. This kind of column could reduce energy consumption and investment costs. The reduced numbers of columns and reboilers could also make a plant more compact, important for offshore facilities.

Acknowledgments

This research was supported by a grant from the Gas Plant R&D Center funded by the Ministry of Land, Transportation and Maritime Affairs (MLTM) of the Korean government.

References

1. Kidnay, A. J.; Parrish W. R. *Fundamentals of Natural Gas Processing*; Taylor and Francis: Baca Raton, 2006.
2. Elliot, D.; Qualls, W. R.; Huang S.; (Roger) Chen J. J. Benefit of Integrating NGL Extraction and LNG Liquefaction Technology. *AIChE Spring National Meeting, 5th topical conference on Natural Gas Utilization (TI) Session 16c-Gas,* **2005**.
3. Sheffield, J. A. Offshore LNG Production – How to Make it Happen, Business Briefing: LNG

Review 2005.

4. Long N. V. D.; Lee M. Y. Improvement of Natural Gas Liquid Recovery Energy Efficiency through Thermally Coupled Distillation Arrangements, *Asia Pac. J. Chem. Eng.*, In Press, **2011**.
5. Manley, D. B. Deethanizer/Depropanizer Sequences with Thermal and Thermo-Mechanical Coupling and Component Distribution. U.S. patent 5,673,571, **1997**.
6. Mak, J. Configurations and Methods for Improved NGL Recovery. U. S. Patent 7,051,552, **2006**.
7. Long, N. V. D.; Lee, M. Y. Reduce Costs and Energy Consumption of Deethanizing and Depropanizing Fractionation Steps in NGL Recovery Process. *Computer Aided Chemical Engineering*, **2011**, *29*, 1658-1662.
8. Manley, D. B. Multiple Effect and Distributive Separation of Isobutane and Normal Butane. U.S. patent 8,806,339, **1998**.
9. Lee, S.; Cha, K.; Park, C.; Lee, C.; Cho, Y.; Yang, Y. The Study on Natural Gas Liquefaction Cycle Development. *ISOPE PACOMS,* Busan, Korea, **2010**.
10. Tamayo-Galván, V. E.; Segovia-Hernández, J. G.; Hernández, S.; Cabrera-Ruiz, J.; Alcántara-Ávila, J. R. Controllability Analysis of Alternate Schemes to Complex Column Arrangements with Thermal Coupling for the Separation of Ternary Mixtures. *Com. Chem. Eng.* **2008**, *32*, 3057-3066.
11. Hern´andez, S.; Jim´enez, A. Design of Energy-Efficient Petlyuk Systems. *Com. Chem. Eng.* **1999**, *23*, 1005-1010.
12. Long, N. V. D.; Lee, S. H.; Lee, M. Y. Design and Optimization of a Dividing Wall Column for Debottlenecking of the Acetic Acid Purification Process. *Chem. Eng. Process.* **2010**, *49*, 825-835.
13. Poth, N.; Brusis, D.; Stichlmair, J. Minimaler Energiebedarf von Trennwandkolonnen. *Chem. Ing. Tech.* **2004**, *76*, 1811–1814.
14. Amminudin, K. A.; Smith, R.; Thong, D. Y. C.; Towler, G. P. Design and Optimization of Fully Thermally Coupled Distillation Columns: Part 1: Preliminary Design and Optimization Methodology. *Trans. IChemE.* **2001**, *79 (Part A)*, 701-715.
15. Box, G. E. P.; Behnken, D. W. Some New Three Level Designs for the Study of Quantitative Variables. *J. Technometrics*, **1960**, *2*, 455-475.
16. Long, N. V. D.; Lee, M. Y. Dividing Wall Column Structure Design using Response Surface Methodology. *Comp. Chem. Eng.* **2012**, *37*, 119-124.
17. Naklie, M. Mobil's Floating LNG Plant. *International Conference & Exhibition on Liquefied Natural Gas 12th*, Perth, Australia, 4-7 May, **1998**.
18. Kerbers, I.; Hartnell, G. A Breakthrough for Floating LNG. *24th World Gas Conference*, Argentina, 5-9 October, **2009**.
19. Shell Development (Australia) Proprietary Limited (Shell), Prelude Floating LNG Project, October, **2009**.
20. Lee, Y. M.; Cho, T. I; Lee, J. H.; Kwon, O. Y. Trends and Technologies in LNG Carriers and Offshore LNG Facilities. *Offshore Technology Conference*, Houston, Texas, U.S.A, 5-8 May **2008**.
21. Aspen Technology. Aspen HYSYS Thermodynamics COM Interface, version number V7.1, January 2009.

Simulation and Optimization of Rectisol Using Revised Thermodynamic Methods

Lili Liu[1], *Fang Cai*[1], ***Jinsheng Sun***[1], *Wenchao Han*[2]
[1]*School of Chemical Engineering and Technology, Tianjin University, China*
[2]*China Special Equipment Inspection and Research Center (CSEI), Beijing, China*

1. Introduction

Rectisol is an economical, eco-friendly and well-proven process for the removal of acid components from raw gases, such as syngas and natural gas [1]. It uses hypothermal methanol as a solvent because it has a high vapor pressure and a strong absorptive capacity for acid gases in the normal operation range of -30 to -100℉(-34.4℃to -73.3℃) [2]. Rectisol technology provides an excellent option for the co-removal of many contaminants to downstream catalysts in one integrated unit. Contaminants including H_2S, COS, CO_2, and organic sulphides, were reduced to trace levels (e.g. H_2S to <0.1 ppm by volume). It is the preferred pathway for almost all units of process industry involving strict gas purification requirements [3]. By 2006, over 70 large enterprises saw different Rectisol units in their plants by patent and technology suppliers like Lurgi and Linde [4,5].

Although flourishing in industry, the description of thermodynamic behavior of the hypothermal process is rarely described, hindering accurate process simulation and further process melioration. According to preliminary trials, almost all possible thermodynamic models in prevailed commercial process simulators, such as ProII with Provision and ASPEN, were found to give no workable results compared with on-site data. In an effort to accurately simulate and improve a typical five-column scheme, we revised the thermodynamic parameters according to industrial data through regression analyses and predictions. The revisions of the thermodynamic model presented in this work perfectly describe the current process and facilitate a process optimization-oriented unit heat integration based on pinch and next step exergy analyses.

2. Simulation of the protytope rectisol process

2.1 *Feed conditions*

The feed gas in the simulation is from a carbon monoxide shift conversion device. The feed condition: flow rate 8160 Kmol/h, temperature 245.2K, pressure 3.63Mpa, The characteristics are listed in Table 1.

Table 1. Basic characteristics of the feed gas

Components	Concentration (mol%)
CH_3OH	5.44E-05
CO_2	0.3229
H_2	0.4576
N_2	2.81E-03
H_2S	9.95E-04
H_2O	3.59E-06
Ar	1.00E-03

CH_4	8.10E-04
CO	0.2138
COS	1.00E-15

2.2 *Process description of the target Rectisol*

The typical process of five-column Rectisol is shown schematically in Fig. 1. It is composed of five blocks: 1) acid gas absorption column (C1), 2) CO_2 regeneration column (C2), 3) H_2S concentration column (C3), 4) methanol regeneration column (C4) and 5) methanol/water separation column (C5).

The raw gas from the upstream carbon monoxide unit is cooled by hypothermal methanol in E1 and then flash liberated in V1 before flowing into C1. The moisture is frozen and solidified for removal, avoiding the adverse blocking effects in downstream equipment. Within this absorption operation, most H_2S and CO_2 are removed, leaving the purified raw gas which is transported to the liquid nitrogen wash unit for further treatment involving cryogenic recovery of carbon monoxide and oxo-gas. Two sidelines are drawn from the upper and lower column sections. The upper side cuts flow to C2, whereas CO_2 is flashed out before being transported to the urea or methanol synthesis unit and the lower to C3. The dissolved gas (except for H_2S) in the bottom product of the lower absorption column is stripped by low pressure N_2. In addition, the methanol solution, rich in H_2S, is preheated before being transported to C4. It is then regenerated through rectification, which is collaboratively powered by bottom reboiler heating and vapor stripping by the top product of C5. H_2S and other trace impurities are recovered at the top of the C4 column. Finally, the lean methanol is refined in C5 and pumped back to absorb raw material gas.

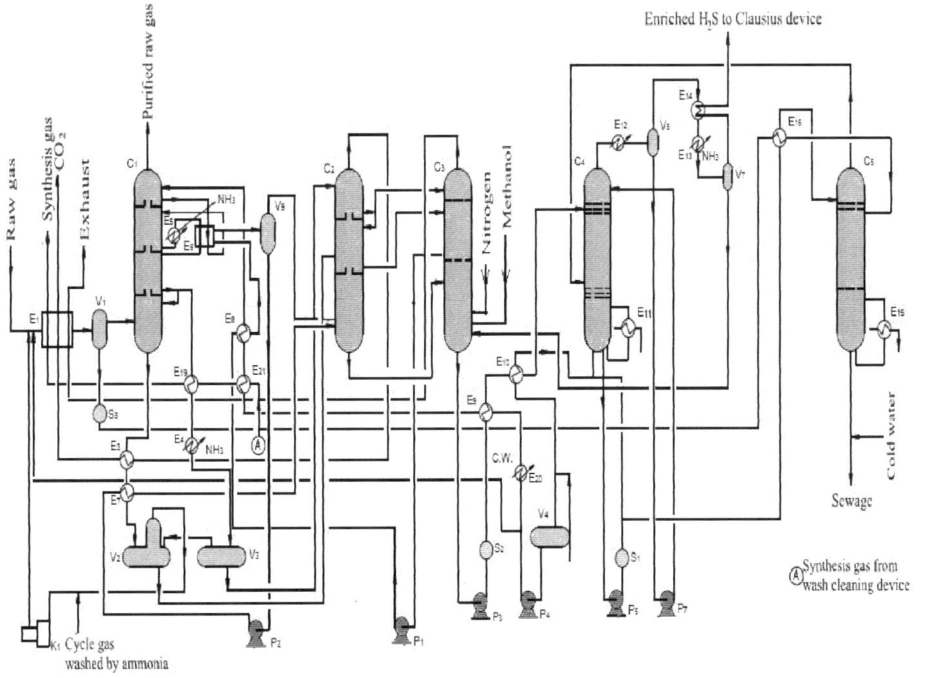

Figure 1. Flow sheet of five units Rectisol

2.3 *Thermodynamic models*

Occasionally, in process-oriented research, there is a specific, non-ideal material system or a special extreme operation condition under which the systemic thermodynamic behavior is far from normal. Similarly, the inbuilt thermodynamic models or simulation methods fail to accurately describe the absorptive process involving methanol, carbon dioxide and hydrogen sulfide under low temperature and high pressure conditions. In order to achieve a veracious simulative process description, revision of the thermodynamic model is inevitable. The convenient and fast approach involves selecting the thermodynamics models which are as close to on-site results as possible, then rectify the parameters of this thermodynamic prototype with field or experimental database.

Before the thermodynamic revision, we split the complex columns into functional blocks for simplification purpose [6, 7]. C1 is the first column in the process and deviates the most thermodynamically. This column was divided into 4 parts: 1) H_2S absorption section (C110), 2) CO_2 minor absorption section (C120), 3) main absorption section (C130) and 4) fine absorption section (C140). The next target, C3, falls into the upper column (C320) and the under column (C310).

Secondly, different models were screened in advance to simulate these blocks or sections suitably, including PSRK [8,9], RK-ASPEN, RKSWS, SR-POLAR, PRWS,
KSMHV2 and PRMHV2. As shown in Table 2, the appropriate thermodynamic models, whose simulative results most closely matched the field data, were adopted in the corresponding correction.

Table 2. Thermodynamics model of every block

Block	Model
C110	adjusted RK-ASPEN
C120	SR-POLAR
C130	SR-POLAR
C140	SR-POLAR
C200	RK-ASPEN
C310	PSRK
C320	RK-ASPEN
C400	PSRK
C500	PRWS

In C110, calculated results by RK-ASPEN illustrated that stage temperature and stream composition of most components, except for CO_2, resembled field data within permissible deviation. Considering this extreme situation, the authoritative data in the literature [10] was, for the first trial, adopted to modify the related parameters (tabulated in Table 3.) in RK-ASPEN model. Apparently, the simulative computation results supported by this modified model, listed in Table 4., verify a satisfactory effect in contrast with their industrial counterparts. Preliminary results are obtained using a previous RK-ASPEN model and final results were modified.

Table 3. Parameters regressed

Parameters	Component i	Component j	Valve
RKAPOL	CO_2		-0.6561249
RKAPOL	CH_4O		0.48309335
RKAKA0	CO_2	CH_4O	-0.4446355
RKAKA1	CO_2	CH_4O	1.74342955
RKAKB0	CO_2	CH_4O	-0.3010129
RKAKB1	CO_2	CH_4O	1.07304549

Table 4. Comparison between the simulation results and field data of steam103 in the C110 column

flow rate (Kmol/h)	field data	preliminary results	final results
CH_3OH	4.377	2.01218	1.7011
CO_2	2528.2345	1985.553	2536.7
H_2	3732.81	3741.108	3742.1
N_2	22.9006	23.09295	23.109
H_2S	0.00037	0.000327	0.0003
H_2O	0.000029	0.003208	0.0025
Ar	8.17532	8.142333	8.162
CH_4	6.5829	6.619009	6.6469
CO	1743.6932	1751.876	1753.3
COS	6.76E-12	3.61E-12	3.64E-12

3. Analysis and comparison of calculation results

According to the preparatory work above, the prototype flow sheet of the Rectisol process, shown in Fig. 2, was up built and simulated. Special attention was paid to comparisons between simulation results and field data, including stage temperature, total stream and main component flow rate.

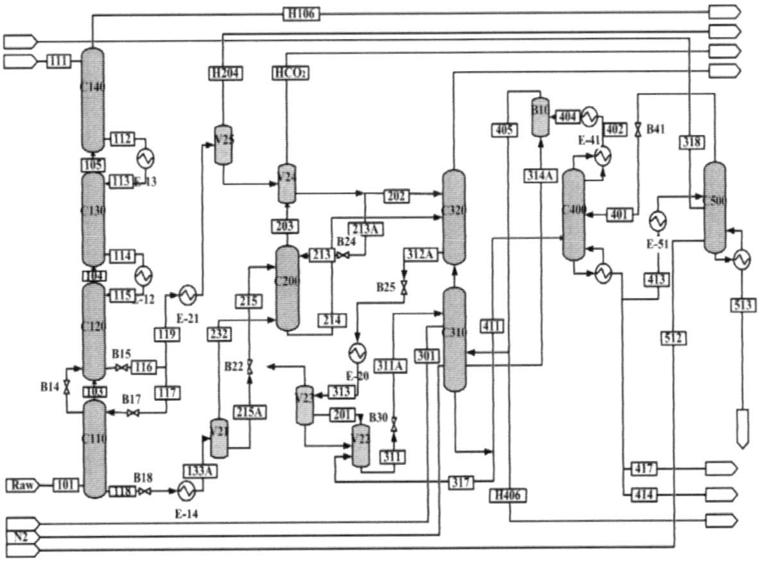

Figure 2. The base-case flow sheet of Rectisol process

Stream 103 in C110, as well as 104 in C120, 105 in C130, 106 in C140, 214 in C2, 312 and tail gas in C3, 401 in C400, 402 in C500, were chosen for this purpose.

Table 5. reveals that the simulation results agree well with the field data, with most errors less than 5%. For specified input feedstock and conditions, the results of simulation based on the revised thermodynamic models are reliable for process analysis and optimization. Their parameters were adjusted using a published database or regressive field correspondence. With continuous addition of more field data to the reservoir, the binary interaction parameters are expected to be more precise and versatile for various feeding conditions.

Table 5. Comparison between simulation and field data

Streams	Items	Simulation	Field data	RE/%
	CO_2 (Kmol/h)	2518.28	2528.24	0.39
103	T (K)	246.3	252.4	2.42
	F (Kmol /h)	8052.72	8046.78	0.07
	CO_2 (Kmol/h)	2016.91	2032.19	0.75
104	T (K)	245.3	249.3	1.6
	F (Kmol /h)	7549.63	7543.58	0.08
	CO_2 (Kmol/h)	1549.83	1548.99	0.05
105	T (K)	246.1	249.1	1.2
	F (Kmol /h)	7080.61	7053.99	0.38
	CO_2 (Kmol/h)	150.3	163.5	8.07
106	T (K)	221.8	232.5	4.6
	F (Kmol /h)	5669.44	5626.65	0.76
	CO_2 (Kmol/h)	318.27	318.57	0.09
203	T (K)	230.6	229.1	0.65
	F (Kmol /h)	325.21	325.43	0.07
	CO_2 (Kmol/h)	1153.06	1147.01	0.53
302	T (K)	258.4	240.3	7.53
	F (Kmol /h)	1605.5	1602.92	0.16
	CO_2 (Kmol/h)	2033.05	2020.8	0.61
Tail gas	T (K)	235.7	216.4	8.92
	F (Kmol /h)	2473.4	2471.01	0.1
	CH_3OH (Kmol/h)	2033.05	2020.8	0.61
311	T (K)	235.7	216.4	8.92
	F (Kmol /h)	2473.4	2471.01	0.1
	CH_3OH (Kmol/h)	103.34	104.67	1.27
401	T (K)	374.7	374.2	0.13
	F (Kmol /h)	107.67	107.68	0.01
	H_2S (Kmol/h)	45.25	45.67	0.92
402	T (K)	318.7	319.3	0.19
	F (Kmol /h)	139.87	139.88	0.01

4. Heat intergration and process optimization

The thermodynamic preparation facilitates systemic checking and analysis for flow sheet defects. The following revision or optimization will aid the energy-saving oriented exergy and enthalpy analyses. The concern here is the connective relationship of the blocks instead of the outer periphery blocks.

4.1 Heat integration

The original heat exchanger network (HEN) of the Rectisol process, as illustrated in Fig. 3 involved six sets of cold utility (CU) distributed in the six pre-cooling locations of five columns. CU is mainly offered through energy and a cost-intensive ammonia refrigeration system, which brings out the significance of more reasonable matching of heat sources and sinks in this scheme. With this in mind, necessary flow sheets were focused on the heat exchanging system, in hope of a more effective suggestion for energy conservation and absorption of acid gas.

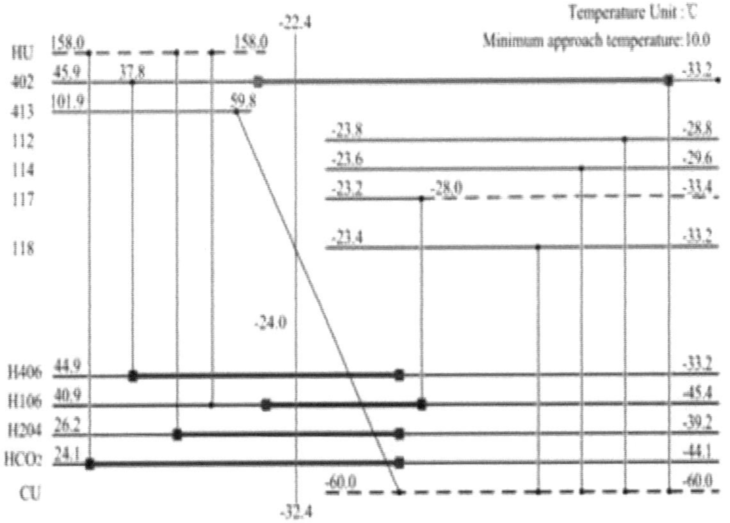

Figure 3. The original HEN

In order to reclaim the cryogenic energy within the system, four cold flows were added to the basis of the original HEN. These cold flows include the following: the separated gas from the overhead of the acid gas absorption column (H106), the flash tank top gas (H204), the Klaus gas (H402) and the flash tank top gas (labeled HCO2). The prefix, 'H', of these four cool streams separates them from the hot streams.

The original HEN was analyzed through pinch analysis [11, 12], using 10℃ as minimum approach temperature. This led to the following findings: 5 heat exchangers transfer heat across the pinch point, whereas the CU is used under the pinch point. These results are contrary to the principles and common senses of pinch analysis.

In order to obtain optimal HEN, the task was broken down into three steps. First, the appropriate minimum approach temperature was selected by thermodynamic and economic consideration. Fig. 4 shows the analysis of the economic condition at different minimum approach temperatures. With the minimum approach temperature increasing, the annual operating costs go up while the total heat transfer area decreases. Considering that the cost of heat-exchange equipment is lower than that of energy consumption, the largest area of heat transfer and the minimum approach temperature of 10℃ were finally adopted. Second, the temperature-enthalpy diagram of the original network and the total portfolio curve were plotted, as shown in Fig. 5 and Fig. 6. The streams, which stayed up and down the pinch, were re-matched in order to fully recover the heat and cold in the system. Finally, the optimal heat exchanger network could be obtained by appropriate adjustments, as shown in Fig. 7.

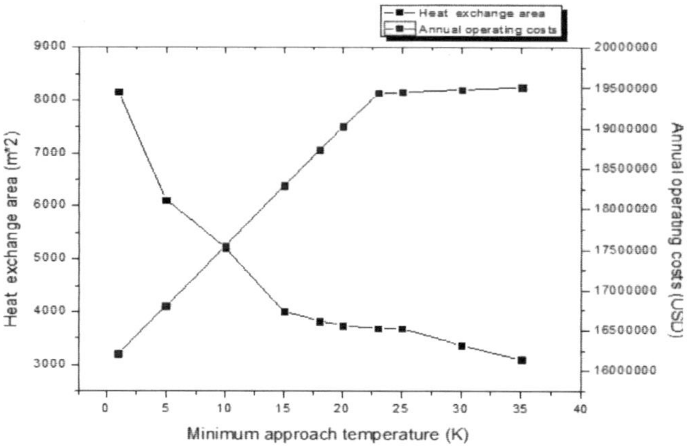

Figure 4. Variation of heat exchange area and annual operating costs due to minimum approach temperature

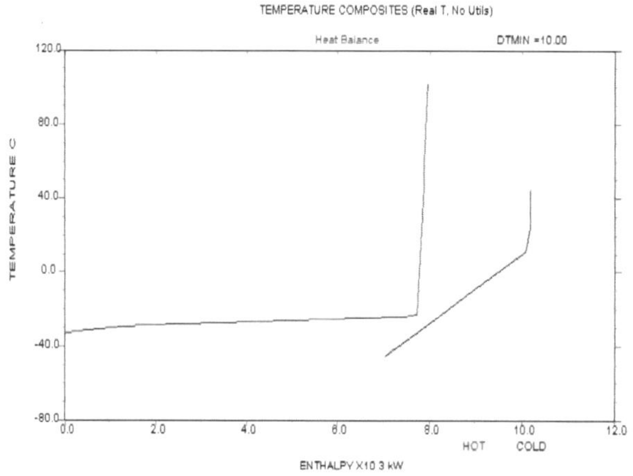

Figure 5. The temperature-enthalpy diagram of HEN at 10℃

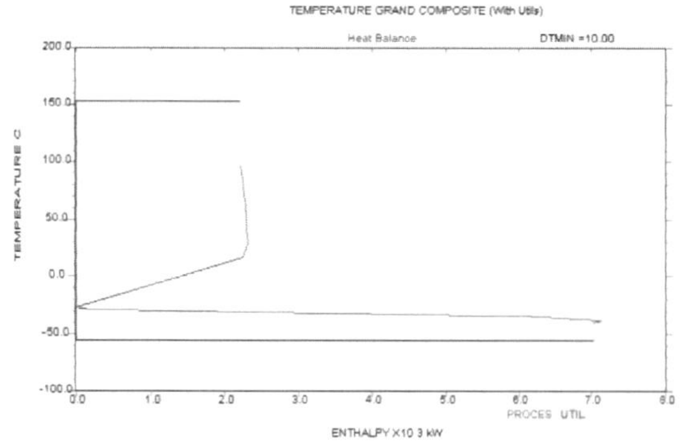

Figure 6. The GCC of HEN at 10℃

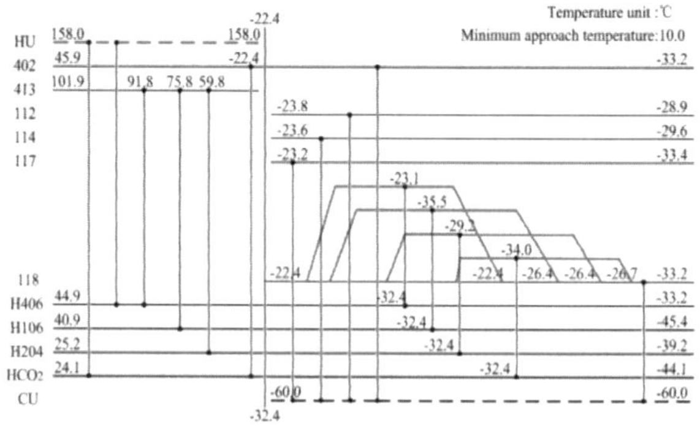

Figure 7. The optimized HEN

The optimized HEN was proven to be superior to the existing prototype network. At the cost of increasing the heat interchanging area of heat exchanger, 327.2 kw hot utilities (HU) and 356.4 kw cold utilities were calculably reclaimed. From an economic perspective, $97,200 per year will be saved by merely adding one-off equipment and a construction investment worth $32,800. It will take only 0.337years to recover the capital by advance estimation under specific stipulation of pinch analyses.

Compared with the original process, the new heat interchanging area increased by 36.36%, and HU and CU consumption were reduced 12.86% and 4.82%, respectively. Additionally, annual operating costs and annual aggregate costs were cut by 5.25% and 5.18%, in the new scheme. Detailed data is shown in the Table 6.

Table 6. Economic comparison between new heat exchanger network and original exchanger network

	Original HEN	New HEN
Number of heat exchanger	11	15
Heat transfer area (m^2)	3999.5	5213.3
Installation costs ($\times 10^3$\$)	104.8	137.6
Heat consumption (KW)	2544.9	2217.7
Cool consumption (KW)	7394.1	7037.7
Hot utility consumption ($\times 10^3$\$/Yr)	982. 6	856. 3
Cold utility consumption ($\times 10^3$\$/Yr)	17536.9	16691 .6
Annual capital cost ($\times 10^3$\$/Yr)	34 .9	45 .9
Annual operating costs ($\times 10^3$\$/Yr)	18519.5	17547. 9
Annual aggregate costs ($\times 10^3$\$/Yr)	18554.5	17593.8

4.2 Process optimization

The exergy efficiency of columns can be evaluated by the exergy analysis method [13] in the Rectisol process. The analytical results, shown in Table 7., indicate that the exergy efficiency of the methanol regeneration column and methanol/water separation column is not satisfactory. The reason is that external energy is added into the two columns. In the methanol regeneration column, cooling water is used for overhead condensing and product cooling. Similarly, the methanol/water separation column consumes external energy to evaporate the column bottom liquid. Therefore, simulative process rearranging is carried out in connection with two columns.

The transformation program of the methanol regeneration column is as follows: stream 411 is preheated from 88.5℃to 98.4℃by one tributary of stream 413, which is cooled from 101.94℃to 91.85℃. In addition, the side stream from the second tray is heated from 91.21℃ to 97 ~ 99℃ by the other tributary of stream 413, then returns to the previous tray. Those of methanol-water separation column are stream 511 and 318, which are heated from 57.9℃ and 55.3℃ to 71.75℃and 125℃, respectively, by wastewater from the bottom. Stream 512 is heated from -28℃ to 12.05℃ by circulating cooling water. After process optimization, annual hot utility consumption will be decreased by approximately 12505.33 kW, 11.5% of that of the unadjusted.

Table 7. Exergy efficiency of columns

Columns	Exergy loss	Exergy Input	Exergy efficiency
C1	4295.669	-2081377	99.894%
C2	1739.021	-452334.55	99.616%
C3	1939.537	-1251213.6	99.845%
C4	9185.14	-459841	98.003%
C5	78487.379	-503578	84.414%

5. Conclusion

Based on published reservoir and field data, a Rectisol process is numerically set up and simulated with the aid of a commercial process simulator. In other words, a secondary development of the program for a first-principles method is developed. Furthermore, HEN and process are optimized by the aid of two main thermodynamic analysis methods: pinch and exergy analysis. The main conclusions are as follows:

(1) The modified thermodynamic model is suitable for simulating the Rectisol process and favorably conformed to the industrial site measurement, with the accurate and reliable parameters regressively obtained.

(2) The original HEN was analyzed to have two unreasonable aspects. The problems were solved by the simulative re-matching streams and by adding four cold flows supported by the adjusted thermodynamic models.

(3) In all, $97,200 per year will be saved at the price of $32,800 one-off investment to improve HEN.

(4) The relatively low exergy efficiency of methanol regeneration column and methanol-water separation column is raised in the optimized system, with annual hot utility consumption decreased by 12505.33 kw.

(5) On the basis of this work, more can be done to improve industrial units or reconfigure more advanced schemes for high energy efficiency.

References

1. Weiss, H. (1988), “Rectisol wash for purification of partial oxidation gases” Gas Separation & Purification, pp.171-176.

2. Olajire, A.A.(2010), “CO2 capture and separation technologies for end-of-pipe applications – A review” Energy, pp. 2610-2628.

3. Sadhukhan, J. Zhao,Y.R. Shah, N. and Brandon, N.P. (2010), “Performance analysis of integrated biomass gasification fuel cell (BGFC) and biomass gasification combined cycle (BGCC) systems” Chemical Engineering Science, pp. 1942-1954.

4. Satoshi, K. Hideyuki, K. Tohru, S. and Kiyohisa, O. (2006), “Photoelectrochemical reduction of carbon dioxide at p-type gallium arsenide and p-type indium phosphide electrodes in methanol” Chem.Eng.J., pp. 227-231.

5. Kaneco, S. Iiba, KJ. Hiei, N.H. Kiyohisa, O. Mizuno, T. and Suzuki, T. (1999), “ Electrochemical reduction of carbon dioxide to ethylene with high Faradaic efficiency at a Cu electrode in CsOH/methanol” Electrochimica Acta, pp. 4701-4706.

6. LIEBMANN, DHOLE K.V.R. and JOBSON, M. (1998), “Integrated Design of a Conventional Crude Oil Distillation Tower Using Pinch Analysis” Chemical Engineering Research and Design, pp.335-347.

7. Watzdorf R.V, Bausa, J. Marquardt, W. (1999), “Shortcut Methods for Nonideal Multicomponent Distillation: 2. Complex Columns” AIChE Journal, pp. 1615-1628.

8. Holderbaum, T. Gmehling, J. (1991), “PSRK: A Group Contribution Equation of State Based on UNIFAC” Fluid Phase Equilibria, pp. 251-265.

9. Fischer, K. Gmehling, J. (1995), “Further development, status and results of the PSRK method for the prediction of vapor-liquid equilibria and gas solubilities” Fluid Phase Equilibria, pp. 1-22.
10. Knapp, H. Döring, R. Oellrich, L. Plocker, U. Prausnitz, J.M. (1982), “Vapor-Liquid Equilibria for Low Boiling Substances” DECHEMA, Frankfurt, Binary Systems:part1.
11. Matijaseviæ, L. Otmaèiæ, H. (2002), “ Energy recovery by pinch technology” Applied thermal Engineering, pp. 477-484.
12. Zhelev, T.K. Semkov K,A. (2004), “ Cleaner flue gas and energy recovery through pinch analysis” Journal of Cleaner production, pp. 165-170.
13. Fawzi, B. Nesreen, J. (2008), “ Exergy analysis of desalination by solar-powered membrane distillation units” Desalination, pp. 27-40.

() Divided Wall Column for Industrial Multi Purpose Use

Daniel Staak[1], Thomas Grutzner[2], Brian Schwegler[2]and Detlef Roederer[2], (1)Department of Process Dynamics and Operation, Technical University of Berlin, Berlin, Germany, (2)Lonza Chemical R&D, Visp, Switzerland

Divided wall columns (DWC) for the distillation of multicomponent mixtures have received much attention in the past 15 years and have experienced a booming development. Although the concept isn't new and originates from the 1930s [1], it took a couple of decades until scientific investigations focused again on heat coupled columns [e.g. 2, 3]. Finally BASF in Ludwigshafen set up the first production scale column in 1985 [4]. The main advantages of DWC are the considerable savings in energy consumption and investment costs. This became increasingly important in times of rising energy costs and increased market competition. Therefore the number of installed units continuously increased since the 80's and today this technique can be considered to be well established in chemical industry.

Since DWC demand higher efforts in equipment design and process control they are predominantly used in continuously operated dedicated production plants. Very little literature can be found where DWC are used in multi-purpose plants. The equipment in those plants, typically batch distillation, has to show a high degree of flexibility – an attribute that hardly seems to fit for divided wall columns.

Lonza set up a multi-purpose DWC on the production site in Visp that fully meets the demands of a steadily changing production. The column is constructed from hastelloy c22 which allows the distillation of corrosive mixtures used in this plant. A flexible inlet enables the feeding on different column segments as well as a flexible outlet allows a side draw from different stages. Also simultaneously feeding on both sides of the dividing wall is possible which allows an operation mode comparable to that of a traditional distillation column This paper reports on the equipment design and the specific technical solutions that had to be implemented in order to meet the harsh specifications of multi purpose production. Particular attention is put on the modelling and simulation of the column and the control scheme that has been applied. The launch and operation of the column is explained in detail and finally the original simulations are compared to real operation data.

[1] A. J. Brugma:*Process and Device for Fractional Distillation of Liquid Mixtures, more Particularly Petroleum, US Patent 2.295.2561942(filed 1936)*

[2] F. B. Petlyuk V. M. Platonov, D. M. Slavinskii:*Thermodynamically Optimal Method for Separating Multicomponent Mixtures*, Int. Chem. Eng. 5,**1965**, 555-561

[3] W. J. Stupin, F. J. Lockhart:*Thermally Coupled Distillation: a Case History*, Chem. Eng. Progr. 88,**1972**, 71-72

[4] BASF company website

(151a) Control of Kaibel Distillation Column by Estimated Compositions From Temperature Measurements

Maryam Ghadrdan, Chemical Engineering Department, Norwegian Univeristy of Science and Technology, Trondheim, Norway, Ivar J. Halvorsen, ICT, SINTEF, Trondheim, Norway, Sigurd Skogestad, Department of Chemical Engineering, Norwegian University of Science and Technology, Trondheim, Norway and Deeptanshu Dwivedi, Deparment of Chemical Engineering, Norwegian University of Science and Technology, Trondheim, Norway

Control of Kaibel Distillation Column by Estimated Compositions from temperature measurements

Maryam Ghadrdan[1], Ivar Halvorsen[2], Sigurd Skogestad[1], Deeptanshu Dwivedi[1]

[1]Department of Chemical Engineering, Norwegian University of Science and Technology, N-7491 Trondheim, Norway, Email:ghadrdan@nt.ntnu.no, skoge@nt.ntnu.no,dwivedi@nt.ntnu.no

[2]SINTEF ICT, Applied Cybernetics, N-7465 Trondheim, Norway, Email:ivar.j.halvorsen@sintef.no

Reliable and accurate measurement of product compositions is one of the important issues in distillation column control. However, direct composition measurements are expensive, not so reliable and there is also a significant delay. Temperature measurements are robust, fast and cheap and representsproven technology in the process industry.

We have used a soft-sensor to estimate the compositions from temperature measurements. Our estimation method (we call it Loss method) is a reformulation of the well-developed Self-optimizing method proposed by Skogestad (2000) to make it adequate for the purpose of estimation (Ghadrdan et al. 2011). We have formulated the Loss method for open-loop and closed-loop estimator. With the term "open-loop" estimator, it is implied that the predicted variables are not used for control purposes. In "closed-loop" the estimates are used as feedback in control loops, and the performance of the closed loop control will depend on the estimator performance. Since the estimator is based on steady-state information, it is important that the behavior in a dynamic system does not degrade the control.

In this work, we are demonstrating our estimation method for a Kaibel distillation arrangement which is a complex case-study. Kaibel distillation arrangement (Figure 1) separates a four-component mixture into pure products and thus replaces a sequence of three conventional distillation columns. There is a potential for significant energy savings, but this requires that the column is operated optimally. The extra operational degrees of freedom in the column are used for the purpose of optimization and to operate the column as close as possible to the minimum energy. The tight integration in Kaibel column makes it challenging to control, compared to the conventional sequence of simple columns. The control performance using the estimated compositions to be controlled in face of various process disturbances will be demonstrated by dynamic simulations.

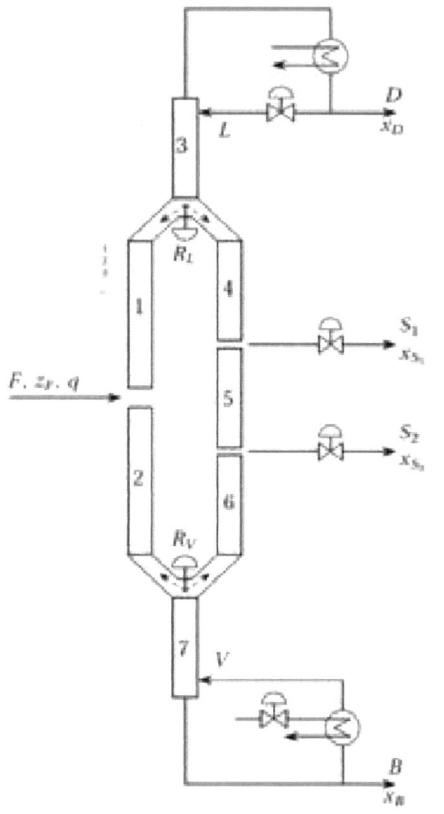

THE GRAPHIC ON THIS PAGE WAS NOT AVAILABLE ON THE E-MEDIA VERSION OF THIS PAPER.

Figure1. Schematic of Kaibel Distillation Column

References

Skogestad, S. (2000).-Plantwide control: the search for theselfoptimizingcontrol structure. J. Proc. Control**10**:487-507

Ghadrdan,M. ,Grimholt, C., Skogestad, S., Halvorsen, I.J.,Lossmethod: Astaticestimatorapplied forproductcompositionestimationfromdistillationcolumntemperatureprofile. AIChE Annual Meeting, (2011), Minneapolis, USA

(151b) Optimal Design of the Integrated CGC and Depropanization System for An Ethylene Plant

Jie Fu, Chuanyu Zhao and Qiang Xu, Dan F. Smith Department of Chemical Engineering, Lamar University, Beaumont, TX

As the most important and valuable facilities, Multistage Compression Systems are generally employed jointly with the Front-End Depropanization section in ethylene plants. In most cases, the Charge Gas Compressor is designed with four or five stages of compression and aftercoolers, which allows the condensation and removal of the heaviest components of the quench tower overheads at the earliest possible time, thus avoiding expending energy on them in later compression. Meanwhile, the low operating temperature of the compressor reduces the tendency for polymerization and fouling in the machine. The Front-End Depropanization section consists of a two tower system: high pressure (HP) column and low pressure (LP) column, with each tower operating at a different pressure. This offers a significant savings in refrigeration and minimizes the fouling problems experienced in a one-tower system.

This study is focused on energy consumption optimization by improved configuration design of a typical integrated CGC-DeC3 system of an ethylene plant. In order to study the integrated system behavior, four stages of compressors with suction drums, Caustic Wash Tower, Liquid Condensate Dryer, HP and LP Depropanizers with reboilers and condensers, C2 Convertor and other related process units are included in the case study model. One innovative design is conducted under the consideration of introducing part of the heavier stream from early stages of compression section directly into the fractionation section. Effluent from the 2nd stage of compression is split into two streams with one sent to the 3rd stage compressor, and the other directed to the low pressure depropanizer, with the intention that saving compressing work on heavy components in following compression, as well as reducing high pressure depropanizer operation load. Key process variables such as split ratio, compression ratio, new feed tray location, column pressure are examined and determined by the optimization while the separation specification and safety requirements are satisfied.

OPTIMIZATION OF A SUPERSONIC EJECTOR COUPLED WITH A CFD ANALYSIS

Amanda Mattos(1), *Karolline Ropelato*(1), *Ricardo Medronho*(2)

(1) ESSS – Engineering Simulation and Scientific Software, Florianopolis/SC, Brazil.
(2) UFRJ – Universidade Federal do Rio de Janeiro, Rio de Janeiro/RJ, Brazil.

ABSTRACT

The present work studied the behavior of ejectors which are widely used in industry as substitutes for mechanical compressors or even for vacuum producers in distillation towers. They drag a secondary fluid through the difference of pressure generated by the entrance of a high pressure motive fluid on the main nozzle which has a combination of converging/diverging ducts. A preliminary study using Computational Fluid Dynamics (CFD) was validated against experimental data. Then the effects of geometry on ejector's performance were investigated. The CFD model was capable to predict well the experimental data. For adequate phenomenon representation, a density based solver was used with a steady-state approach for time prediction, ideal gas treatment for the working fluid and a two-dimensional (2D) approximation. High velocities were achieved after the primary nozzle entrance and a supersonic flow took place. For better use of computational resources, an independency test for the flow pattern was carried out by varying mesh refinement. As a second part of this study, an optimization analysis was developed. In this step, the goal was to achieve the best geometric configuration for an ejector by varying three different geometrical parameters simultaneously with a fixed operating condition and maximizing efficiency of the device. The parameters varied were: primary nozzle inlet diameter where the high pressure motive fluid enters, the length throat which is responsible for flow stabilization and finally the inlet diameter of the mixture chamber which is the entrance of the secondary fluid. The device efficiency was evaluated by the entrainment ratio which is the rate between the entrance of secondary fluid and of motive fluid. In order to create an initial population for the problem, the Central Composite Designs (CCD) model was used as a design of experiments. Moreover for the optimization itself the Non-dominated Sorting Genetic Algorithm II (NSGA-II) was employed with the software ModeFrontier 4.3 from ESTECO. The results for the optimization cases show a direct effect on the ejector performance of all geometrical parameters tested. Increases on the selected parameters provide larger entrainment ratios under the same operating condition. The same analysis also shows that highest sensibility on the device efficiency was achieved by variations on the primary nozzle diameter. Small increases on this diameter generated a much higher drag of secondary fluid.

Keywords: CFD, ejector, supersonic, optimization.

INTRODUCTION

Computational fluid dynamics can be an important tool for the representation of some complex phenomena that occur within certain equipments. This technique solves the conservation equations of interest by using numerical methods. The present work used CFD to validate a representative model for the flow within ejectors. The flow in this type of equipment is characterized by compressibility effects and supersonic velocities which makes it a difficult problem to solve.

Ejectors are devices that can be taken as interesting alternatives to mechanical vacuum pumps or conventional compressors used in refrigeration cycles. The main reason for that is because ejectors use a high pressure steam as a motive fluid, which is usually available in chemical plants. Therefore resulting in electricity saving since this type of energy is required for operating the conventional equipments. Other advantages that can be cited are lack of moving parts and ease of maintenance. The ejector's operating principle is based on the transformation of potential energy into kinetic energy by its favorable geometry, promoting the suction of a secondary fluid through a side entrance and the compression of the mixture over the device, as it is shown by Sriveerakul *et al.* [1].

The literature presents a lot of work in this area. The first studies on this apparatus were based on the one-dimensional theory developed by Keenan and Neumann [2] but it was a limited model that did not take into account the effects of ejector's geometry on its performance. Over time there has been progress on the computational resources and it became possible to use CFD technique in modeling the flow inside ejectors [1,3,4,5]. In the 90´s, Riffat and Omer [3] used a commercial CFD package to analyze the efficiency of a methanol driven ejector. This work shows that the prediction method for the flow inside the ejector must be carefully chosen, especially regarding variations in density and compressibility effects of the working fluid, since sudden changes in pressure due to expansions and restrictions in the geometry are quite common in this equipment. They also tested different primary nozzles and observed how they dramatically affect the performance of equipment, leading to believe that this is an important geometric parameter in the performance analysis of the utensil in question. Even though the work of Riffat and Omer [3] was not validated with experimental data, Sriveerakul *et al.* [1] verifies that it is possible to use computational fluid dynamics as an efficient tool for predicting rate and pressure drag of ejectors. In Sriveerakul *et al.* [1] study a CFD model was developed and validated by the comparison between calculated and experimental values for the pressure profile, demonstrating the accuracy of the model.

For better understanding of an ejector performance, Chunnanond *et al.* [4] showed that the total efficiency of an ejector can be expressed as a function of mixing efficiency or entrainment and the relationship between energy input and energy output. The most desirable scenario for an ejector is to work under such conditions that provide the maximum drag and maintain the highest possible pressure in the discharge. This can be analyzed by the parameter called entrainment ratio which is the rate of secondary fluid drag to the motive fluid insertion. Moreover Sriveerakul *et al.* [5] studies the flow inside ejectors among different geometries, analyzing the effect of certain geometric parameters (such as primary nozzle diameter, suction chamber inlet diameter and throat length) on the equipment's entrainment ratio.

The main idea of this work is to use CFD as a tool for predicting ejector's performance and be able to optimize its efficiency by varying simultaneously three different geometric parameters under a fixed operating condition.

MODELING APPROACH

In order to apply computational fluid dynamics, it was used the commercial code Fluent v 13.0 from Ansys. This technique was further integrated with the optimization software ModeFrontier 4.3 from ESTECO. The developed model had some approximations which together created a representative tool and at the same time had a reduced computational cost so that the optimization process could simulate a greater number of cases in a viable time range.

The first approximation in the model was to simplify the geometry using symmetry along the longitudinal axis along with a 2D slice of the computational domain. The three-dimensional effect was evaluated by the use of an axisymmetric option. This hypothesis affects considerably the cost of computational resources required for a computer simulation, therefore being of great importance for the optimization process development. Prior study of Pianthong *et al.* [6] made comparisons between pressure profiles of a three dimensional and an axisymmetric model (ASXM) showing that the ASXM representation is good enough to give accurate results and the 3D model is not essential for the same analysis type. Figure 1 illustrates the fluid domain of both ejector's entrances which are the primary nozzle at the center and the suction nozzle at the sideway (a small thickness was implemented for viewing purpose of the 2D representation but there was no element on the Z direction).

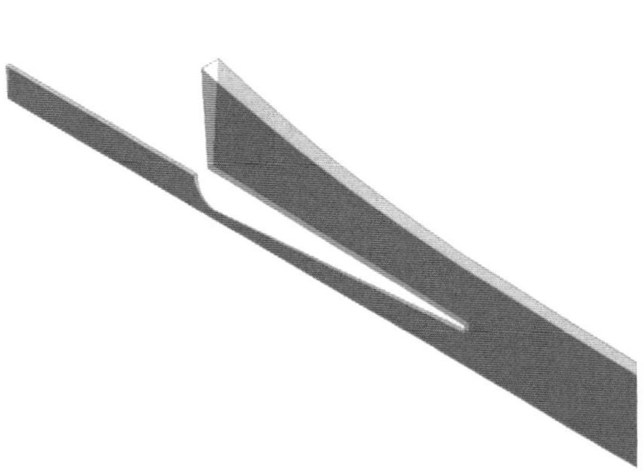

Figure 1: Two dimension representation of a steam ejector.

Selection of ejector's mesh was accomplished by an independency test which had by main objective to identify the variations on pressure and velocity profiles under different levels of refinement. The ideal mesh should be able to capture the phenomena of interest and at the same time have lower computational cost; both of these characteristics were achieved with the mesh independency test.

The first developed mesh had approximately 110000 nodes. Thereafter were created another 3 meshes with 40000 nodes, 13000 nodes and 8000 nodes (all of them maintaining the same refinement pattern from the first one). The comparison between meshes is in Figure 2. Taking into account a decreasing order of refinement (starting from mesh 110 to mesh 8), it is possible to observe that pressure and velocity profiles at the ejector's center line are maintained approximately constant until the 8000 node mesh is tested. This indicates that meshes with less refinement than the 13000 node mesh are not accurate because they are not be able to capture the same profiles obtained with more refined meshes.

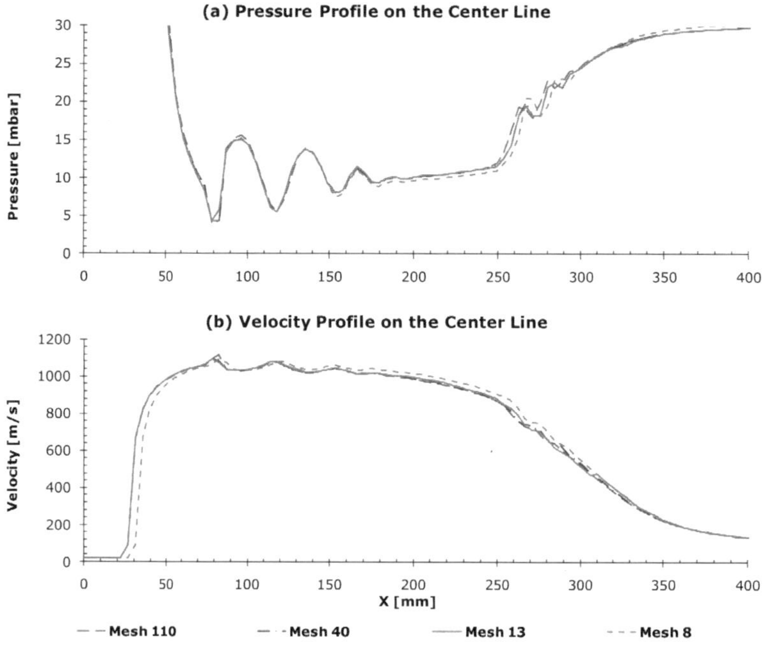

Figure 2: Comparison between results of various meshes considering: (a) pressure profile on the center line and (b) velocity profile on the center line.

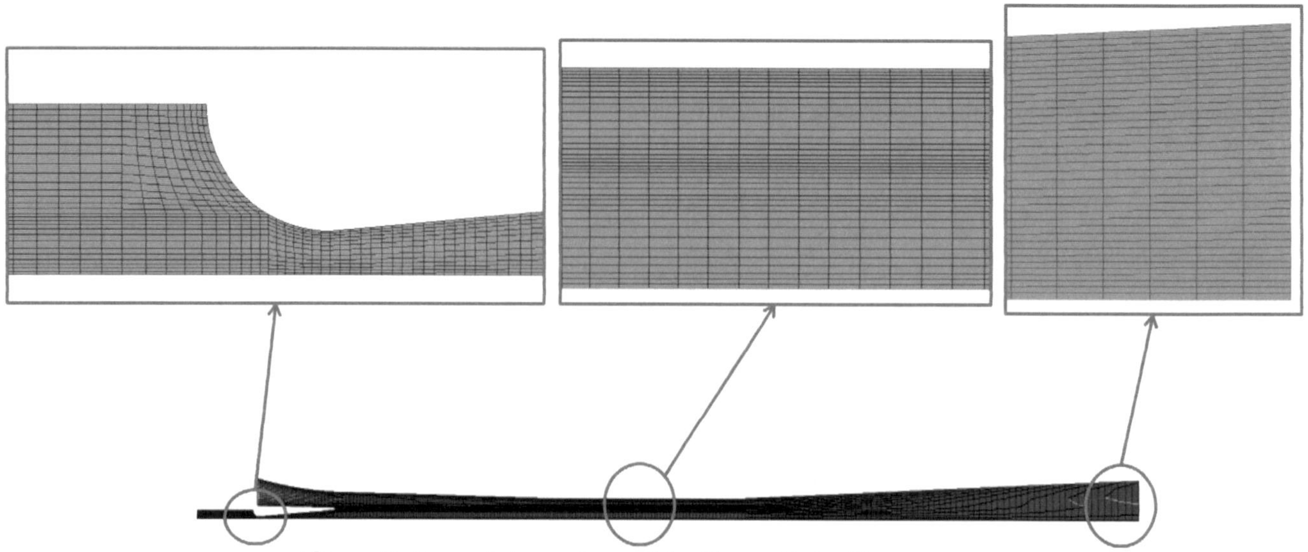

Figure 3: Hexaedric mesh used for the optimization process.

Figure 3 shows the quadrilateral mesh chosen for the next steps of this study and its main refinement characteristics.

Furthermore, a steady state time approach was adopted along with a "*coupled-implicit*" density based solver [7]. For turbulence, it was chosen the *k-εpsilon* model which relies on the Boussinesq hypothesis and utilizes an equation averaging for the Reynolds stress tensor [8]. The advantage of this turbulence model is the relatively low computational cost associated with the determination of the turbulent viscosity. Previous studies show that this is a suitable model to represent the evolution of pressure profiles along steam ejectors [9,10].

The working fluid adopted for the simulations was water vapour. All properties of this fluid were maintained constant throughout the simulation except density which was evaluated with an ideal gas relation. Previous studies show that the ideal gas treatment provides similar results to a more complex alternative such as real gas treatment [11]. Beyond that, the boundary conditions were related to a fixed operating condition by using the working fluid saturated pressures on both inlets and outlet.

The ejector's efficiency was evaluated by the following equation:

$$RM = \frac{mass\ flow\ of\ suction\ fluid\ [\frac{kg}{s}]}{mass\ flor\ of\ motive\ fluid\ [\frac{kg}{s}]} \quad (1).$$

This equation was used in the optimization process which had by main objective to maximize the entrainment ratio. A fixed operating condition was used and three parameters were varied simultaneously as Figure 4 illustrates.

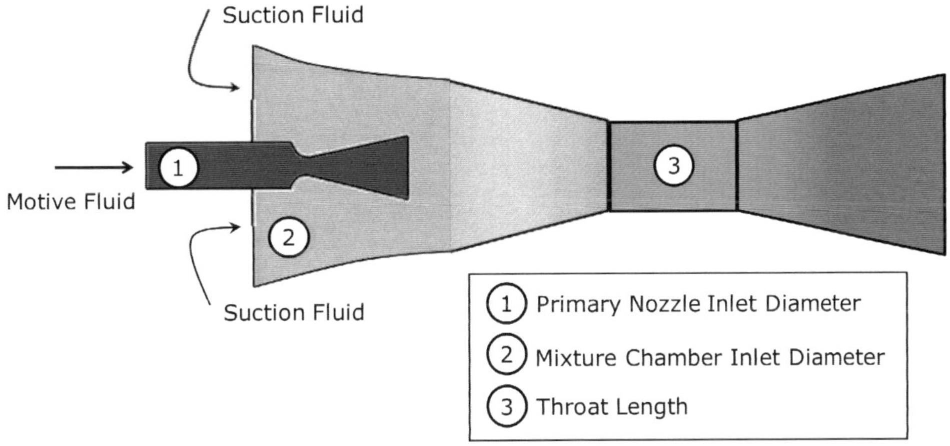

Figure 4: Esquematic view of the three geometrical parameters varied on the optimization process.

For the optimization development, initially it was used the technique of DoE (Design of Experiments) in order to create an initial population for the problem. Design of experiments is a methodology that maximizes the knowledge gained from experimental data [12]. It provides a strong tool to design and analyze experiments or simulations, eliminating redundant observations and reducing the time and resources to make the simulations. Hence, it allows the maximum extraction of information from a limited number of test runs. In this case, Central Composite Designs (CDD) was selected as the DoE model since it allows studying the effect of the variables on the response, as well as the effects of interactions between variables [12].

Moreover the Non-dominated Sorting Genetic Algorithm II (NSGA-II) was employed on the second phase of the optimization process. This algorithm guarantees the diversity and spread of solutions without use of sharing parameters [12]. Also it implements elitism for multi-objective search and the constraint handling method does not make use of penalty parameters.

All dimensions for the ejector's geometry and its operating conditions were retrieved from Sriveerakul *et al.* [1] which provides also the experimental results for entrainment ratio and pressure profile along the ejector's wall. The conditions for the optimization process can be summarized on Table 1.

Table 1: Summary of variables optimization range.

	Optimization Range
Primary Nozzle Inlet Diameter	6 - 8 [mm]
Mixture Chamber Inlet Diameter	19 - 29 [mm]
Throat Length	19 - 114 [mm]

RESULTS AND DISCUSSION

The CFD results for pressure and velocity profiles of both fluids on the center line are shown in Figure 5. It characterizes a typical steam ejector's operating principle [1] where high-pressure steam (point 1) is injected into the primary nozzle and expands at low speed due to the Venturi phenomenon. This results in a dramatic increase in the velocity of the primary fluid (reaching supersonic speeds) and consequent reduction in its pressure. This low pressure that is generated at the primary nozzle exit (point 2) provides a pressure differential, in relation to the suction chamber inlet, such that a secondary fluid is dragged by side entrance (point 3). The motive fluid, which is now at high speed enters the mixing chamber and is in contact with the suction fluid. The contact between the two fluids characterizes the slowing of motive fluid while the secondary fluid's velocity is increased to reach sonic values.

At the end of mixing chamber, the two fluids are completely mixtured (point 4) and static pressure stays approximately constant until reaching the diffuser region. The mixture reaches a maximum speed and a minimal pressure on the throat and due to a higher pressure region in the output of ejector, a compression effect emerges and there is a sudden decrease in flow velocity (from supersonic to subsonic).

Finally, the mixture proceeds by the diffuser where it suffers a slight compression effect and is then discharged in the output of equipment with an intermediate pressure between the two inlet pressures (between the motive fluid pressure and the suction fluid pressure).

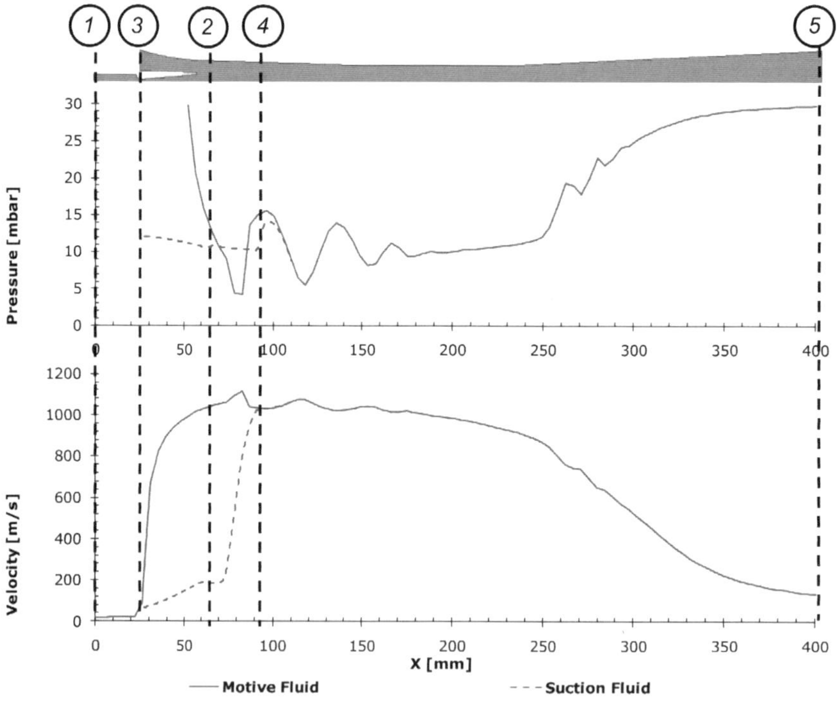

Figure 5: Schematic view of a steam ejector and its velocity/pressure profiles.

The comparison between experimental data (provided in literature by Sriveerakul *et al.* [1]) and calculated data from the CFD model are in Figure 6. The simulation was conducted when the ejector was operated at primary fluid saturated temperature (TP) of 120 °C, secondary fluid saturated temperature (TS) of 10°C and downstream pressure (PC) of 30 mbar.

The calculated pressure profile on the steam ejector wall shows a good agreement with the experimental values (with errors under 10%). Therefore this validates the CFD model for this operating condition and confirms that the developed methodology is able to represent the phenomenon inside the ejector. This is also verified when comparing entrainment ratio values which for experimental data is 0.53 and for calculated data resulted in 0.54 showing less than 2% of error.

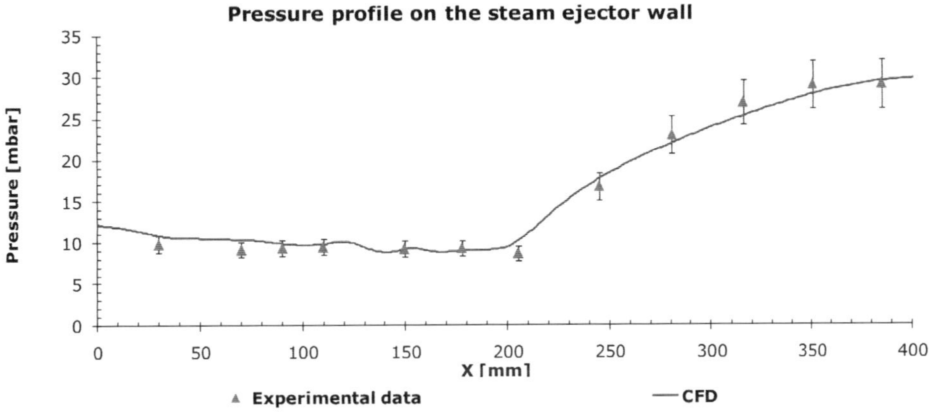

Figure 6: Comparison with calculated and experimental data for pressure profile along the ejector wall.

The simulations regarding the optimization part can be summarized by Figure 7 which shows the history of this process with 150 test runs and their respective calculated entrainment ratio. The optimal point is shown at the end of the graphic. It provides an increase of 25% over the original ejector's entrainment ratio just by varying three geometric details in the equipment's geometry.

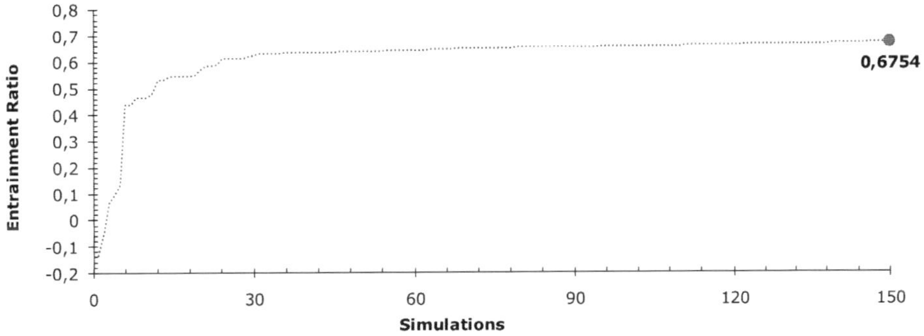

Figure 7: History of optimization process.

A sensibility analysis was also made and the results are shown in Figure 8. In this type of graph it is possible to estimate relationships between output and input values, in this case it indicates how the selected geometric parameters affect the entrainment ratio. The height of the bars is called "*Effect size*" and shows the strength of this relationship [13]. All effect sizes were greater than zero attesting a direct relationship with the entrainment ratio.

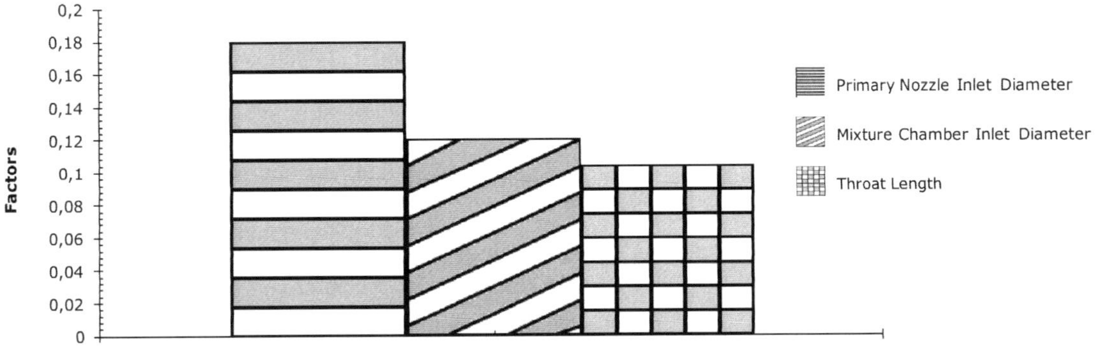

Figure 8: Effects of the optimization variables over the objective function.

Supporting the sensibility analysis, the significance values obtained for variable's effect size are summarized in Table 2. The low values of significance mean that the previous results (the values of "*Effect Size*") are probably true [13] and the effect parameter is reliable.

Table 2: Summary of calculated significances for optimization variables.

	Significance
Primary Nozzle Inlet Diameter	0.0051
Mixture Chamber Inlet Diameter	0.0346
Throat Length	0.0160

Accordingly, the primary nozzle inlet diameter shows a direct effect on the ejector's performance along with the highest influence on its efficiency, meaning that small increases on this dimension will result in larger entrainment ratios.

CONCLUSIONS

It was verified that the CFD method is an efficient tool to predict the pressure profile along the ejector's wall and also the equipment's expected behavior. The comparison with experimental values for entrainment ratio shows the accuracy of the developed model. Beyond this, CFD provides more information regarding the flow in the interior of the apparatus such as velocity profiles and visualization of oblique shock waves which are difficult phenomena to capture on experiments.

The integration between CFD method and optimization software provide an economic way to design optimal ejector's geometry without the need of producing hundreds of prototypes. The optimization process helps to understand which geometric variables affect the equipment's performance and the strength of this influence. Further studies can be done in this area regarding more geometric parameters variations or even changes in the ejector's operating condition.

REFERENCES

[1] T. Sriveerakul, S. Aphornratana, K. Chunnanond. Performance prediction of steam ejector using computational fluid dynamics: Part I. Validation of the CFD results. International Journal of Thermal Sciences 46 (2007) 812-822.

[2] J.H. Keenan, E.P. Neumann. A simple air ejector. ASME Journal of Applied Mechanics 64 (1942) 75–81.

[3] S.B. Riffat, S.A. Omer. CFD modeling and experimental investigation of an ejector refrigeration system using methanol as the working fluid. International Journal of Energy Reservation 25 (2001) 115–128.

[4] K. Chunnanond, S. Aphornratana. Ejectors: applications in refrigeration technology. Renewable and Sustainable Energy Reviews 8 (2004) 129–155.

[5] T. Sriveerakul, S. Aphornratana, K. Chunnanond. Performance prediction of steam ejector using computational fluid dynamics: Part II. Validation of the CFD results. International Journal of Thermal Sciences 46 (2007) 812-822.

[6] K. Pianthong, W. Seehanam, M. Behnia, T. Sriveerakul, S. Aphornratana. Investigation and improvement of ejector refrigeration system using computational fluid dynamics technique. Energy Conversion and Management 48 (2007) 2556–2564.

[7] ANSYS FLUENT Theory Guide Release 13.0 (2010).

[8] B. Launder, D. Spalding. The numerical computation of turbulent flows. Computer Methods in Applied Mechanics and Engineering (1974) 3:269e89.

[9] X. Wang, J. Dong. Numerical study on the performances of steam-jet vacuum pump at different operating conditions. Vacuum 84 (2010) 1341e1346.

[10] A. Hemidi, F. Henry, S. Leclaire, J. Seynhaeve, Y. Bartosiewicz. CFD analysis of a supersonic air ejector. Part I: Experimental validation of single-phase and two-phase operation. Applied Thermal Engineering 29 (2009) 1523–1531.

[11] S. Aphornratana. Theoretical and experimental investigation of a combine ejector-absorption refrigerator. PhD thesis, University of Sheffield, UK, (1994).

[12] ModeFRONTIER™4.3 User Manual

[13] D. Montgomery, G. Runger; Applied Statistics and Probability for Engineers 3rd Edition (2002).

METHYLCHLOROSILANE DISTILLATION SYSTEM: COMPARISON OF EXISTING PROCESSES AND NOVEL PARALLEL DOUBLE-EFFECT COUNTERPARTS BY SIMULATION

*Lili Liu[1], Hong Gao[1],Shuo Shi[1], **Jinsheng Sun[1]** ,Shaojing Wang[2], Shuo Jiang[1], Peng Wu[1], Yanzhen Liu[1],Fan Wang[1], Fang Cai[1], Tingting Ma[1]*

1 School of Chemical Engineering and Technology, Tianjin University, Tianjin, P. R. China
2 Tianjin Yilongchuangjia Technology Development Co., Ltd., Tianjin, P. R. China

1. Introduction

Organic silicon materials, mostly polymer form of several chloro-silicane monomers, have been considered 'gourmet powder' of industry all the time. With the finding of their excellent features under extreme conditions, such as sound anti-ageing and isolation at 1000℃and elasticity at -110℃, these materials attract increasing attention for their overwhelming concentration of silicon over carbon in the earth' s crust. Under this circumstance, organic silicon materials are taking over more and more important roles in almost all aspect of modern industry: chemical materials, electronic, aviation, construction, textile, and medicine etc.

Among all the units operations for organic silicon producing progress, monomer distillation for product purification is the key procedure basically affecting the quality of corresponding polymers. Organic silicon monomers include methylchlorosilanes, phenylchlorosilanes and vinylchlorosilanes etc [1]. Methylchlorosilanes, including methyldichlorosilane (MeH), trimethylchlorosilane (Me3), methyltrichlorosilane (Me1), account for over 90% (mass percentage) of the total Methylchlorosilanes, and meanwhile, dimethylchlorosilane (Me2) are the most common product targets of monomer distillation. This process, according to on-site statistics, consumes energy occupying about 70 percent of the whole monomer plant, meaning that any cut of heat supply though technical improvement will be significant for the whole factory.

In China, most producers are suffering average unit energy consumption of 3.5 ton steam per ton monomer product. This situation induces the main concern of this paper is to probe a solution to this by stimulatively discussing 3 novel paralleled heat coupling energy saving processes under the same thermodynamic system.

To show the extent of energy saving, related comparative simulations for the three counterpart traditional schemes of the three double-effect ones were carried out

2. Modella schedule

2.1 Crude monomer feedstock

Crude methylchlorosilane are converted from chloromethane and silicon in the presence of catalysts at high temperature. The normal boiling points and a typical composition in mass percentage of the feeding mixture are listed in Table 1[2, 3].

Table 1 indicates that the boiling points of the components are sufficiently close that Raoult's law must be taken into account. This implies high challenges for monomer distillation when requiring high purities of the distillates. Besides, the binary azeotropic of silicon tetrachloride and trimethylchlorosilane, forms an extra component to be removed.

Table 1. Components of crude methylchlorosilane monomer

component	boiling point at 1atm/℃	mass fraction%
LB		1%
Methyldichlorosilane (MeH)	41.8	4%
silicon tetrachloride (SiCl4)	57.6	0.5%
trimethylchlorosilane (Me3)	57.9	2.5%
Methyltrichlorosilane (Me1)	66.4	11%
dimethyldichlorosilane(Me2)	70.2	80%
HB		1%

The most valuable monomer in organosilicon components are dimethyldichlorosilane(Me2), methyltrichlorosilane(Me1), trimethylchlorosilane (Me3), and methyldichlorosilane (MeH) and so on, in descending order according to their importance and percentage composition. The purification re -uirments on organosilicon components is showed in Table 2[1].

Table 2. Purification demand on organosilicon

monomer	purification demand
Me1H	99.5%
Me3	99.5%
Me1	99.5%
Me2	99.95%

2.2 Flowsheet description for current organic silicon monomer distillation

The current distillation flowsheets fall into 5 types operated under ordinary pressure: 7 columns, 8 columns 9 columns and 10 columns forward/reverse, with almost all of them beginning with a heavies removal column to separate the mixture with the unsteady substances of high boiling points. In the schemes with 9 or more columns, a high boiling point column fed by residue of heavies removal column may allow the upstream heavies removal column to operate at lower bottom temperature keeping the instable heavies from thermal cracking to release more unknown substances which would undermine the sequent separations. The remaining useful components are reclaimed in the high boiling point column distillate, which returns to feed heavies removal column.

Considering the prevailing and overwhelming status of 9-column scheme in China, as shown in Figure 1, the main concern of this work focuses on the improvement to this prototype. The second light ends column in this scheme drive Me3, MeH and other light ends to the top distillate with Me1 and Me2 to the residue. The light ends distillate flows to light removal column, MeH column, azeotrope removal column and Me3 column for two products and two wastes. The Me1 and Me2 mixture is partly separated in Me1 column with pure Me1 in the distillate and about 10% Me1 retained in the residue along with

Me2. This bottom product is pumped to Me2 column where Me2 yields through a vapor side cut adjacent to bottom and the distillate with the same composition as the Me1 column feed circulated back to Me1 column feedstock. The modicum of Me2 column residue may be pumped back to the heavies removal column. The 2 recycles between the heavies removal column and high boiling point column as well as Me1 and Me2 columns lower the operative difficulties of the columns with only one process target to assure for each one of them.

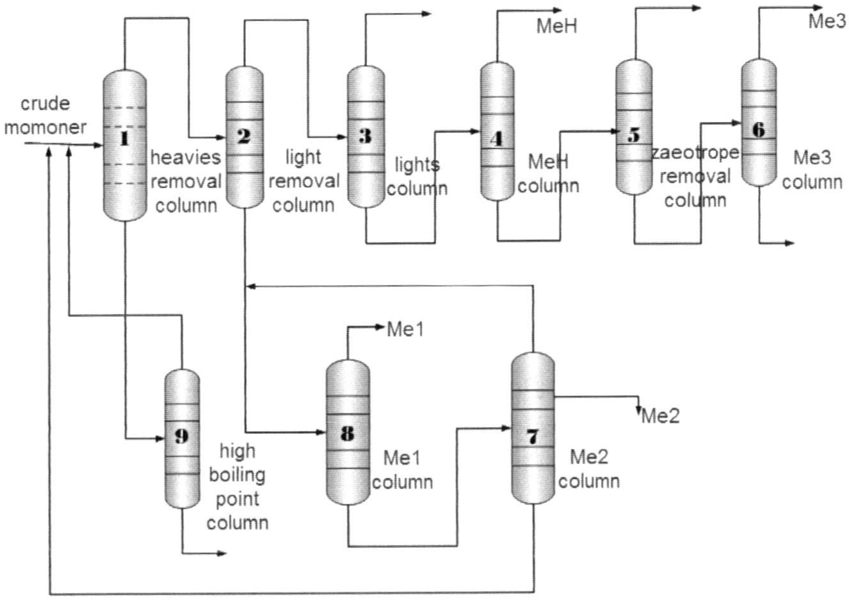

Figure 1. The prevailing 9-column scheme

2.3 Double-effect or heat integrated distillation

Nowadays, the increasing scarcity and high-price of fuel calls for more energy-efficient distillation methods to cut energy requirement. Among the solutions without extra rotating machines powered by electricity or other power producers, double-effect or heat integrated distillation has been the most predilection for the majority of researches [4, 5, 6, 7].

One of the schemes with this effective thermal utilization is the feed splitting (FS) manner of double-effect distillation, in which the feed stream is split into approximately equal two parts and sent to the two separated distillation towers running a different operation pressure, as illustrated in Figure 2a. This difference in pressure of the cascading columns causes the between adjacent condensing vapor form the high pressured column and the boiling liquid from the bottom of the low pressured counterpart. In this way, heat from the high pressure condensing vapor is transferred to the low pressure column to boil the liquid there, power the operation of the heat receiving column. Therefore, the heat duties associated simultaneously with condensation and boiling are reportedly equal to the heat duty for single effect with the same total feedstock and separation requirement divided by the number of effect. That is:

$$Q_{effect} \approx Q/N_{effect} \quad (1)$$

Where, Q_{effect} is the heat duty required to power the double effect scheme, Q the heat duty for single effect control and N_{effect} the number of effect.

Obviously Equation (1) does not consider the influence of the variation of the relative volatility between the components with the change of pressure. For instance, ascending operation pressure may result in descending relative volatilities causing subsequently larger reflux ratio for the same processing target as those at relatively low pressures. Undoubtedly, this will be responsible for higher heat loads of the rebolier of the high pressured column, encumbering the Q_{effect}. In a revised version, Equation (1) is rewritten to be:

$$Q_{effect} \approx C \cdot Q / N_{effect} \tag{2}$$

Where, C is the correction parameters, which is the function of pressure, reflecting the impact of it on relative volatility through reflux ratio and heat requirement. Or Equation (2) can be just approximately simplified, when quick decision is acquired, to be:

$$Q_{effect} \approx \frac{v_{atom}}{v_{press}} \cdot Q \Big/ N_{effect} \tag{3}$$

Where, subscripts "*press*" and "*atom*" stand respectively for relative volatility at pressured and atmospheric condition.

In addition to feed splitting (FS) scheme, as described above, there are other two variations for double-effect distillations. For these configurations, as shown in figure 1b and 1c, the common characteristics of them is entire feed stream sent to the first column. In the light split/forward heat integration (LSF) configuration (Figure 2b), the feed is pumped to the prepositive high-pressure column (HPC). About half of the light key component is removed in the distillate at high purity. The bottom product, which contains the remainder of the target component, is fed to the low-pressure column (LPC). In this case, the heat integration is in the direction of mass flow. For the other variation in figure 2c, referred to as the light split/reverse heat integration (LSR) configuration, the feed is sent to the prepositive low-pressure column. Also, about half of the light key component is removed in the distillate, with the bottom product pumped and sent to the high-pressure column. Herein, the heat integration is in the reverse direction of mass flow.

The different zenith pressure or temperature for the 3 schemes, determined by their features, together with the subsequent influence on degree of operative difficulty, are probably decisive in their population in the plants. The FS is rarely encountered because of, if light and heavy key components are required to be fully separated, the high temperature at the HPC bottom through the pressurization for reasonable temperature difference between HPC top and LPC bottom. This operation condition calls for a heat utility with a higher temperature-level than that at the HPC bottom. Unlike FS, the LSF HPC is just demanded the purity of the half light key component product at the top of the first column, living the other half at the bottom. This operation lowers the temperature, according to the effect of boiling point depression, of the HPC bottom and in turn that of the matching heat utility. More favorably, the coherent direction of heat integration and mass flow makes starting of the unit smoother and easily controllable. These advantages make LSF the most popularized scheme. As for LSR, it benefits from the lowest temperature, among the columns 3 schemes, at the bottom of the first column through the dual actions of low pressure and boiling point lowering. This low temperature-level matches a reduced operation pressure of the post-positional pressured column, namely the submit pressure of the whole scheme, compared with the other schemes. The highest temperature occurs at the bottom of the pressured column falling between that of FS and LSF cousins. In simulation comparisons, the LSR, when used in methanol and water distillation, requires less energy than LSF and FS correspondences. The shortcoming blocks it from being the most popularized is the relatively complicated starting procedure and longer stabilization time.

From the descriptions hereinabove, and considering the cascade connected features of organic silicone distillation as well as the importance and difficulty of the starting procedure, the novel heat integrated schemes are developed on the basis of LSF pattern.

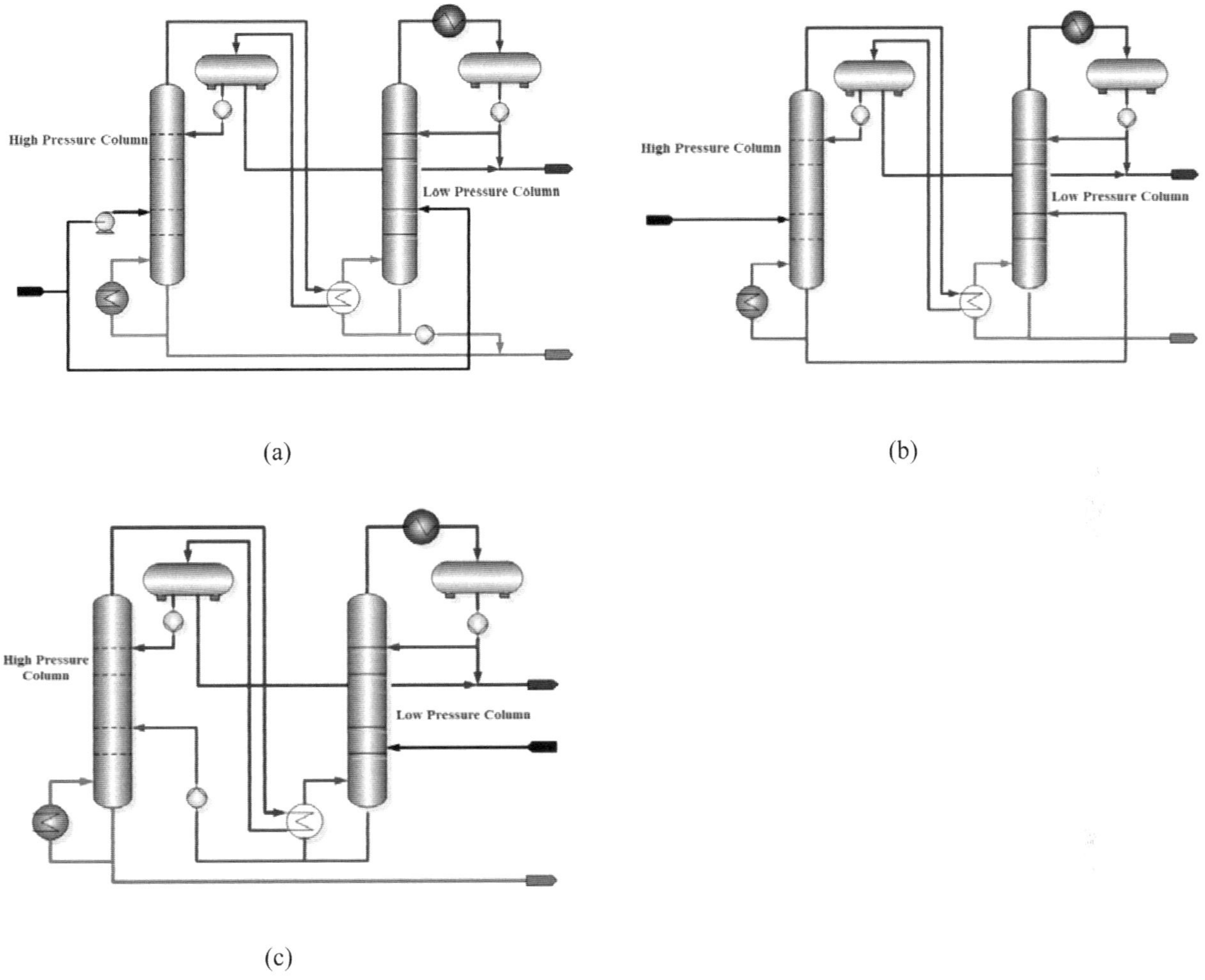

Figure 2. Three configurations of double-effect (FS, LSF, LSR) with high pressured columns always the left

2.4 Novel heat integrated processes

The simulation research in advance revealed that, although industrial reflux ratio data of heavies removal column ranges from 3 to 8, the heat of condensation numerically matched with the total heat load of the other 7 columns of low bottom temperature varying between 75-90℃, except for the high boiling point column. This validates the potential of schematic improvement through heat integration. Thanks to the compression characteristics of the main components in the top vapor, only a slight pressure increase of the operation pressure results in enough difference in temperature for the achievement of the double effect distillations, which could be shown in Figure 3.

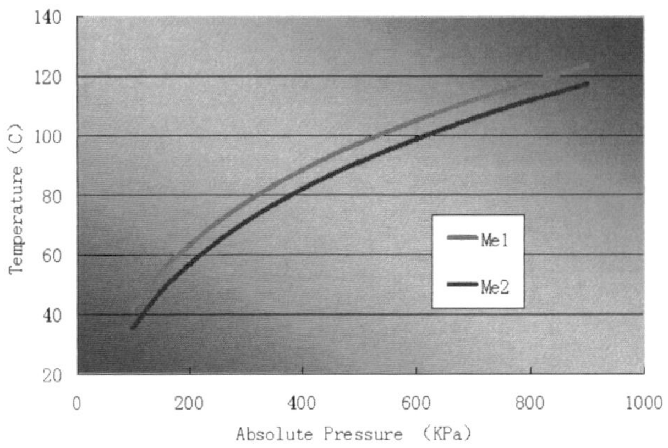

Figure 3. The compression characteristics or thermodynamic equilibrium curves of the main components

To apply the thermal integration to the 9-column scheme, the flowsheets are transformed to 3 manners, as shown in Figure 4~Figure 6[8]. The operation pressure of heavies removal column rises to 360 driving the top temperature to 114℃ with the bottom between 125℃ to 135℃. The hot top vapor distributes into tributaries to reboilers, as heating medium, of low removal column, lights removal column, Me1H column, azeotrope removal column, Me3 column, Me2 column. Me1 column is also pressured to 280 with the top temperature 105℃. The top vapor of Me1 is also induced to partly heat Me2. After that, the hot vapors are cooled to reflux temperature in the reboilers, with their superfluous heat in correspondent special tributaries to be cooled in air coolers. The reflux ratio of heavies removal column is 6.25 for sufficient heat supply. The details can be seen in Figure 4.

When Me1 column, still operated under normal pressure, is also heated by one tributary of the heavies removal column top as shown in Figure 5, another 1-parallel-double-effect process in 9-column scheme appears with only 2 columns powered by steam, other than the Figure 3 counterpart, which has 3 steam reboliers. Same reflux ratio of heavies removal column is adopted as in the case of 2-parallel-double-effect process in 9-column scheme.

Another schematic transformation occurs when Me1 and M2 column work for cascade operation, in which the two columns are theoretically considered to be one column. This is actually a 1-parallel-double-effect process in 8-column scheme which is shown in Figure 6.

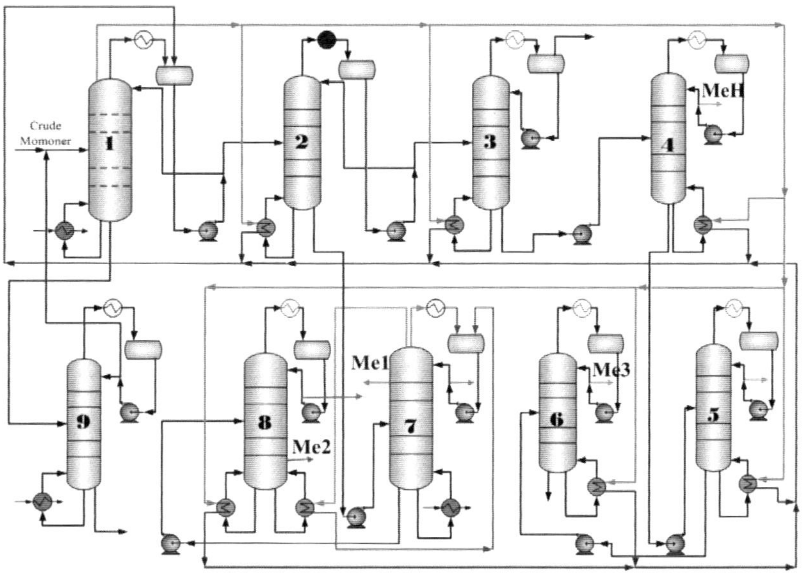

Figure 4. 2-parallel-double-effect process in 9-column scheme:
1— heavies removal column; 2—low removal column; 3—lights removal column; 4—MeH Column; 5—azeotrope removal column; 6—Me3 column; 7—Me1 column; 8—Me2 column; 9—high boiling point column

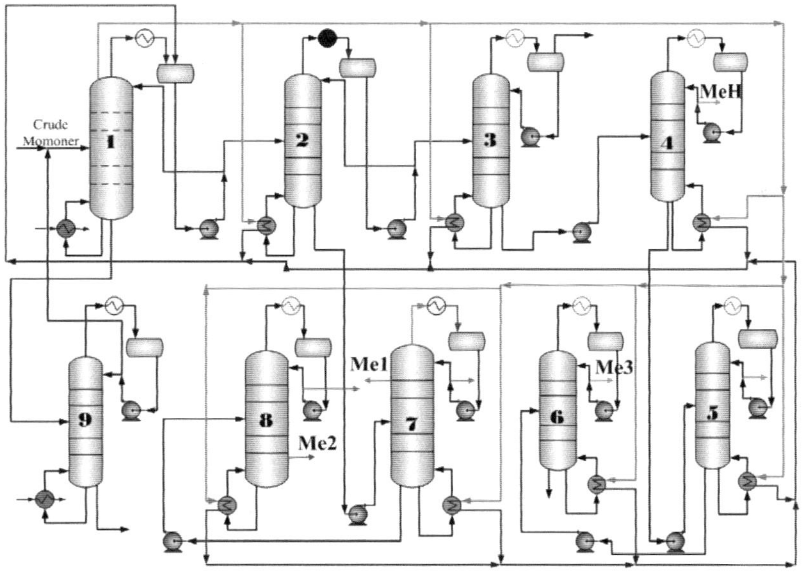

Figure 5. 1-parallel-double-effect process in 9-column scheme:
1—heavies removal column; 2—low removal column; 3—lights removal column; 4—MeH Column; 5—azeotrope removal column; 6—Me3 column; 7—Me1 column; 8—Me2 column; 9—high boiling point column

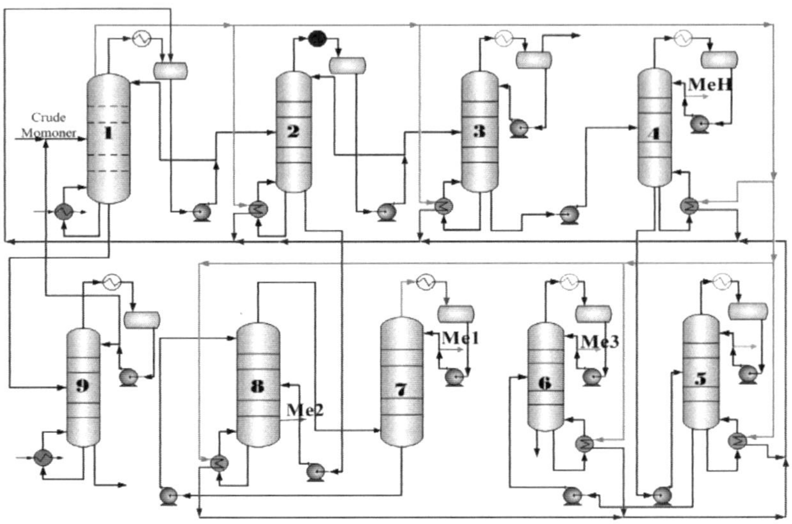

Figure 6 1-parallel-double-effect process in 8-column scheme:
1—heavies removal column; 2—low removal column; 3—lights removal column; 4—MeH Column; 5—azeotrope removal column; 6—Me3 column; 7—Me1 column; 8—Me2 column; 9—high boiling point column

2.5 Comparative simulation and results

The total heat loads of the steam reboliers in the 3 thermal integrated schemes are listed in Table 3. To compare specific energy consumption of the 3 double-effect options with their normal counterparts, the steam requirements of unit pure monomers are summarized in Table 4 from process simulation under same yields and purities.

Table 3. Comparison of heat duty of the processes

process	heat duty of condenser	heat duty of reboiler
9 dual-effect column(2)	-43.306	42.464
9 dual-effect column(1)	-43.141	42.474
8 dual-effect column(1)	-50.771	44.380

According to Table 4, the specific energy consumption is almost halved in comparison with the original schemes without thermal integration. Moreover, the reflux ratio of the heavies removal column is between the industrial of 3-8, still with a fairly large amount of heat carried by monomer vapor removed by air cooling, indicating further potential to reduce energy use.

Table 4. Comparison of steam consumption per ton monomer product in terms of steam requirements

process	Heat integrated scheme	Normal Scheme
9 dual-effect column(2)	1.7490	3.625
9 dual-effect column(1)	1.7494	3.625
8 dual-effect column(1)	1.8279	3.498

3. Conclusions

Significant results are obtained through simulative discuss of the highly energy integrated schematic transformation of the prevailed normal options by the aid of parallel double-effect distillation. This specification, derive from the well known and proven thermal integrative pattern, differs from its prototype by that the heat supplier column serves not only the one downstream column, but also a serious of them. This special system can also embed another double-effect subsystem in it, like what we do in 2-parallel-double-effect process in 9-column scheme. With only 2 or 3 reboliers powered by steam, the novel options halve the specific energy consumption of the total monomer yield, which usually occupies two thirds of the total energy requirement of the monomer factory. More importantly, the new schemes allow further energy cut when lower reflux ratio of the heavies removal column is practically permitted.

On the other hand, air cooling is allowed, even in the tropical and subtropical regions, to the pressured columns for condensation, reducing the usage of circulating cooling water, as well as its evaporation loss and pumping power.

The operation is also under easy control. That is, except for the flowsheet watershed low removal column, the other columns just have one operation target respectively. Besides, the processes are of highly security with the zenith pressures of all the three systems just up to 360Kpa, still in the range of normal pressure.

References

1. P. Budnikov P, (1960), *Glass and Ceramics*, June.
2. H.h. Zhang, (2003), *New chemical materials*, October, pp. 1-3.
3. L. L. Tian, J. J. Wang, X. P. Gu. (2004), *Modern chemical industry*, December, pp. 23-26.
4. E. Fiedler, G. Grossman, et al. (Eds.), (2002), "Ullmann's Encyclopedia of Industrial Chemistry" *Wiley-VCH Verlag, GmbH & Co.*.
5. M. Harvey, (1993), *IMTOF*, London.
6. T. Chiang, W.L. Luyben., (1983), *Ind. Eng. Chem. Process. Des. Dev.*, February, pp. 175-179.
7. A.P. Douglas a, A.F.A. Hoadley., (2006), *Applied Thermal Engineering*, pp. 338–349.
8. J.S. Sun, P. Wu, H. Gao, Y.H. Wang, Y.X. Su, D. Gao, J. Li, (2010), *CN Patent 101798322*, June, pp. 67-75.

A Novel Resilient Process Design of Propylene Hydrogenation Purification by Catalytic Distillation

Yi Liu [1,2*], Xiaodan Gao [2,], Dongfeng Li [1], Sam Mannan [2]**

([1]Beijing Research Institute of Chemical Industry, SINOPEC Cop., Beijing 100013, China

[2]Mary Kay O'Connor Process Safety Center, Texas A&M University, College Station, Texas 77843-3122, USA)

Abstract A novel catalytic distillation process to convert methyl acetylene and propadiene (MAPD) into propylene with a high selectivity is proposed in a propylene catalytic distillation column. In the improved process, MAPD are separated from propylene, enriched at the bottom of the stripping section of the propylene fractionation column and hydrogenated into propylene simultaneously. Compared to conventional hydrogenation process followed by distillation unit, the propylene content in the hydrogenation section is decreased from 90% to as low as 2%, therefore the side reaction is significantly prevented. By Aspen Plus simulation, the propylene yield increases by 2.1%~4.5% with a hydrogenation selectivity increasing significantly from 40%~60% to 99%. The integrated process utilizes the reaction heat as part of the reboiler heat source, resulting in an energy saved by 2.5%~5.6%. Besides, the novel process shows a high hydrogenation capacity and a high anti-deactivation ability to extend the catalyst regeneration cycle by 3~4 times. It makes that the new process is especially suitable to deal with the C3 fraction from heavy cracking feedstock or shorter residence time cracking furnace.

Keywords: Catalytic distillation; propylene; hydrogenation; deactivation; purification; naphtha steam pyrolysis

* Corresponding author. Tel.:+1 979 739 6964;
** Corresponding author. Tel.:+1 979 862 6910;
E-mail addresses: ly99@tsinghua.org.cn (Dr. Y. Liu), xiaodan_gao08@mail.che.tamu.edu (Dr. X.Gao).
Y. Liu is now working at Mary Kay O'Conner Process Safety Center, Texas A&M Univ., TX, USA;

1. Introduction

Naphtha steam pyrolysis in ethylene plants is one of the main sources of propylene product. There are hydrogen, hydrocarbons such as C_1, C_2, propane, propylene, methyl acetylene (MA), propadiene (PD), C_{4+} and so on in the cracking products. Based on different separation processes with tail-end hydrogenation, e.g. sequential, front-end deethanizing and front-end depropanizing procedures, C_3 fraction flows are usually separated from the overhead of depropanizing column or the bottom-draw of deethanizing column [1,2]. Then the C_3 fraction is hydrogenated before being sent to propylene distillation column (PDC), where propane and MAPD are separated from propylene and enriched at the stripping section, while chemical-grade or polymer-grade propylene is obtained from overhead or side-draw effluent (Fig.1) [1,2].

Methyl acetylene and propadiene (MAPD) usually account for 1~5 mol% of the C_3 fraction and they are highly undesirable as contaminants in propylene product even at a low content of 5 ppm in many processes [3,4], e.g. catalytic oxidation to propylene oxide, polymerization etc. Catalytic hydrogenation is generally employed to eliminate MAPD and therefore to increase the propylene yield as reactions (1) and (2). However, current industrial hydrogenation units are generally applying adiabatic fixed bed reactors and the selectivity of MAPD converted to propylene generally only reaches 40%~60% [3-9]. The reason is that propylene in the C_3 fraction is at a high content of 80%~95% and it can also be overly hydrogenated into propane as reaction (3), reducing the propylene yield [3,4,6].

$$CH_3-C\equiv CH+H_2\rightarrow CH_3-CH=CH_2 \tag{1}$$

$$CH_2=C=CH_2+H_2\rightarrow CH_3-CH=CH_2 \tag{2}$$

$$CH_3-CH=CH_2+H_2\rightarrow CH_3-CH_2-CH_3 \tag{3}$$

Much work to improve the catalyst has been done to reduce the side reaction, however only a little improvement is achieved [3-17]. Besides, due to the gradual densification of cracking feedstock or application of shorter residence time cracking furnace, the MAPD content in C_3 fraction is usually as high as 5%~8%, resulting in much more reaction heat and larger burden to improve hydrogenation capacity.

In this article, a novel catalytic distillation process is proposed to improve the selectivity of MAPD conversion and to reduce the burden of hydrogenation units. By Aspen Plus simulation, the new process is proved to be effective to increase the propylene yield, to save energy, to decrease the influence of deactivation on the process, to have a good adaptability for heavy cracking feedstock, and even to eliminate heat transfer equipments.

2. Novel catalytic distillation processes

2.1. Conventional processes

The propylene separation units are shown in Fig.1. C_3 fraction stream, from the overhead of depropanizing column or the bottom-draw of de-ethanizing column, is sent to PDC after hydrogenation. Propane and MAPD are separated from propylene and drawn out as the bottom product, while chemical-grade or polymer-grade propylene is obtained from overhead or side-draw effluent.

Conventional hydrogenation processes are generally performed in adiabatic fixed bed reactors. Nearly 80%~95% of C_3 fraction feedstock is propylene, while MAPD accounts for circa 1%~5%. The MAPD should be reduced to 5~100 ppm at the outlet of the reactor (The outlet MAPD concentration can be a little above 5 ppm, because PDC can separate MAPD form propylene further), so catalysts with high selectivity and optimized process conditions are critical to increase the propylene yield [3-9]. Besides, due to the gradual applications of heavy cracking feedstock or short residence time cracking furnace, the MAPD content in C_3 fraction is usually as high as 5%~8%, resulting in a larger amount of reaction heat. Complicated multi-stage reactors with intercooling and flash drum are usually employed; otherwise, for a single-stage reactor (Fig.2), most of effluents should be recycled to inlet stream after being cooled, leading to a lower hydrogenation capacity [3-6]. Moreover, regeneration should be adopted to deal with the deactivation. Therefore the process is much more complex and expensive.

If the hydrogenation is combined in the depropanizing column, namely catalytic distillation process [10-17] as shown in Fig.3(a), the C_3 fraction will be separated from C_4 fraction and hydrogenated simultaneously before being drawn out from overhead. The reaction heat can be removed directly as the latent heat of vaporization, therefore the temperature is easy to control, while the heat transfer equipment and reactors are saved. However, this process cannot take full advantage of catalytic distillation in separating unfavorable materials that can cause side reactions. MAPD is not separated from propylene, whose content is still 80%~95% in the hydrogenation section, similar to the conventional hydrogenation in fixed bed reactors (Fig.1 and Fig.2). There is only a little improvement in the hydrogenation selectivity due to the better temperature control. Besides, the catalyst deactivation has an obvious worse effect on the catalytic distillation process than conventional hydrogenation in fixed bed reactors.

2.2. A novel catalytic distillation process

The boiling point of MAPD is higher than those of propylene and propane, so it can potentially be separated from propylene by normal distillation. In fact, it has been the primary method to separate MAPD in the early stage of ethylene industry [4,5,18,19]. But the fraction of MAPD should be kept below 40 wt% due to the hazard of decomposition or polymerization explosion. To reduce the MAPD concentration in the kettle, part of propylene should be taken out from the bottom, which results in a large loss of propylene [4,18]. Furthermore, the MAPD content should be less than 2% when the bottom-draw product is used as a fuel such as liquefied petroleum gas [2]. The MAPD will also increase the heater coking rate if the bottom-draw stream is recycled to the cracking heater [5]. It seems not a good choice to separate MAPD only using PDC. It is also the reason that conventional catalytic hydrogenation is mostly used to eliminate MAPD.

The novel process takes both the advantages of distillation and catalytic hydrogenation technology, to install a catalytic distillation structure at the bottom of propylene distillation column as shown in Fig.3 (b) and Fig.4, which is called as propylene catalytic distillation column (PCDC). C_3 fraction, from the overhead of depropanizing column or the bottom-draw of deethanizing column, is directly (not hydrogenated) sent to PCDC, where propylene is obtained as overhead or side-draw while MAPD is enriched in the stripper section and the kettle. Hydrogen is fed in below the catalytic distillation structures and flows upwards. MAPD, contained in the liquid and the rising gas, is hydrogenated on the catalytic distillation structures. In the new process, the propylene content at the hydrogenation section is decreased from 90% to as low as 2%, which is good for improving the selectivity of MAPD conversion; on the other hand, the MAPD content can be as large as 0.2%~2% or even more, not need to be as low as 5-100 ppm [3-6], so it is efficient to increase the reaction rate and selectivity.

Aspen Plus simulation was performed on a PCDC with 160 trays (120 stages equivalent) and a 2-m-high catalytic distillation structure placed under the 160th tray. C_3 fraction with a MAPD content of 3.5% and a propylene content of 92.8% was fed on the 85th tray at a flow rate of about 22 t/h. Hydrogen containing 5% methane was fed below the catalytic distillation structures at a flow rate of 54 kg/h. Then MAPD was converted into propylene with a high selectivity of 99%. The gas drawn after two-stage partial condensers was 45 kg/h, containing circa 26 kg/h propylene. The condensate was totally refluxed to the PCDC. The bottom-draw rate was 0.81 t/h, where the contents of propylene and MAPD were only 1.8% and 0.27% respectively. Propylene was drawn out from 5th tray at a rate of 21.16 t/h, where MAPD was traceless while the contents of propylene, propane, hydrogen, methane and ethane were 99.61%, 0.34%, 3 ppm, 34 ppm and 0.047% respectively (Tab.1). The propylene product met the standard of polymer-grade propylene with a yield of 0.43 t/h more than that of the conventional hydrogenation process, about 2.1% of total output.

The reaction heat can be easily removed by the latent heat of vaporization in catalytic distillation process. In other words, the energy consumption is reduced; especially the heat transfer equipment and cooling medium are saved. Compared to conventional process, the energy saving is about 2.5%; for heavy cracking feedstock, it can even be as high as 5%~6% (Tab.1).

The vapor/liquid loads, temperature and composition profiles of the PCDC in the novel process are shown in Fig.5. Compared with the conventional process (hydrogenation in fixed bed reactors), there are nearly no change in vapor-liquid loads and only a slight difference in the temperature profiles, similar to those in the concentration profiles of propylene. The reboiler and condenser temperatures change little due to only a few changes in the composition of products, which is beneficial for modification of existing plant. The MAPD content in the feeding tray is higher; however, it decreases quickly nearby the kettle due to catalytic hydrogenation, and it is even lower than the conventional case in the bottom-draw product while the MAPD content in the overhead is lower than 5 ppm. The maximum MAPD content is near 1%, far lower than 40%, so the hazard of explosion is controlled.

2.3. A modified catalytic distillation process

In the catalytic distillation unit, the catalytic distillation structure is installed in the column and cannot be easily regenerated. It is a little difficult for the application of the novel catalytic distillation process. A modified catalytic distillation process is shown as Fig.6. The same feedstock is adopted. However, the catalytic hydrogenation unit is placed outside, which looks like a backpack. The 160th tray is equipped with a special downcomer. The downcomer should be blinded to hold the liquid, so the liquid can be drawn out into the outside hydrogenation reactor while vapor can go upwards through the tray. Hydrogen is introduced into the reactor, where MAPD is converted into propylene with a single pass conversion of 50% and a high selectivity of 99%. The yield of propylene, drawn out from 5th tray, is 21.16 t/h, same to that before modification (section 2.2). The catalysts in the outside reactor can be easily regenerated to deal with the deactivation. Therefore, the modified catalytic distillation process is much more flexible.

3. Performances of new processes

For the conventional hydrogenation process, both the deactivation of catalyst and increasing of the MAPD content in C_3 fraction will leave a high concentration of MAPD unconverted in the effluent. The novel catalytic distillation process, especially the modified process, shows a high anti-deactivation ability and a good performance to remove the reaction heat, so it displays a high capacity of hydrogenation.

3.1. Anti-deactivation behavior

To simplify the calculation, the hydrogenation process is assumed as first-order reaction and isothermal [4,7]. For the single-stage hydrogenation reactor as shown in Fig.2, the flow rate and MAPD concentration of the feedstock stream are assumed as Q and C. To maintain a low MAPD concentration in the inlet stream of the reactor, a recycle stream with a flow rate of Q_c is mixed with the feedstock. The MAPD concentration at reactor inlet and outlet are labeled as C_i and C_o, while the flow rate of stream into the reactor and the residence time is assumed to be Q_r and τ, so

$$C_o = C_i \mathrm{e}^{-k\tau}. \tag{4}$$

$$Q_r C_i = QC + Q_c C_o \tag{5}$$

Usually, $C_o \ll C_i$, $Q_c < Q_r$, so

$$C_i \approx QC/Q_r \tag{6}$$

When the catalytic activity reduces to a ($0<a<1$), the inlet and outlet MAPD concentration (C_i' and C_o') can be expressed as

$$C_o' = C_i' e^{-ak\tau}. \tag{7}$$

$$C_i' \approx C_i. \tag{8}$$

The magnification (M) of the outlet MAPD concentration after deactivation is

$$M = C_o'/C_o \approx e^{(1-a)k\tau} = (C_i/C_o)^{(1-a)}. \tag{9}$$

As for the novel catalytic distillation process, especially the modified process, the distillation and hydrogenation are coupled, and MAPD is enriched in the kettle, which accelerates the reaction rate, thus the deactivation can be weakened. Generally, in the conventional hydrogenation process, $C_i = 1\% \sim 2\%$, $C_o = 100 \sim 300$ ppm, while in the novel catalytic distillation process, the single pass conversion of MAPD is about 50%, so $C_1 = 2C_2$. If the catalyst activity drops to half, then

$$M = \sqrt{C_i/C_o} = 8 \sim 12 \quad \text{(Conventional process),} \tag{10}$$

$$M \approx 2.4 \quad \text{(Novel catalytic distillation process).} \tag{11}$$

According to the results by Aspen Plus simulation, the MAPD content in the effluent of conventional hydrogenation reactor should be controlled as low as 200 ppm to insure the MAPD content in the kettle of PDC to be 0.38%. When the catalytic activity drops to 50%, the MAPD content in the effluent of reactor will be over 2000 ppm, leading to a 10-time rise of the MAPD content in the kettle to 3.8%. As for the novel process, the MAPD content in the bottom-draw product can be as low as 0.27% when the single pass conversion of MAPD in the backpack-like hydrogenation reactor is 50%. If the catalytic activity drops to 50%, the MAPD content in the bottom-draw product will rise to 0.66% (Tab.1), only increased by 1.4 times, matching the result of equation (11) exactly. The effect of catalyst deactivation on the exit MAPD content is greatly weakened in the new process, which indirectly extends the catalyst regeneration cycle by 3~4 times.

3.2. High hydrogenation capacity

The new processes also show a good adaptability for a significant increase of MAPD in C_3 fraction due to the good ability to remove reaction heat.

For the conventional process, if the MAPD content in C_3 fraction rises from 3.5% to 7%, the flow rate Q_r should be doubled based on equation (6) to maintain a low inlet MAPD concentration C_i, which will shorten the residence time τ to be nearly $\tau/2$, and the exit MAPD concentration C_o of hydrogenation reactor will be increased by 9 times (Tab.1).

For the catalytic distillation process, the MAPD content increases to be two-fold when the MAPD content in the feedstock increased from 3.5% to 7%. The MAPD content in the inlet stream is still lower than 1% referring to the results in Fig.5(d), so the influence of reaction heat on the reactor configuration is not obvious (the outlet stream of the reactor does not need to be recycled to the inlet stream). According to the results by Aspen Plus simulation, the MAPD content in the outlet stream just rises from 0.22% to 0.45% (the MAPD content in the kettle rises from 0.27% to 0.55%) for the new process, only increased by 1 time, which is much smaller than that for the conventional process (Tab.1). So the effect of the feedstock on the hydrogenation process is less serious in the new process.

4. Application for heavy cracking feedstock

The novel catalytic distillation processes, especially the modified process, have shown a high selectivity of MAPD conversion, a good energy saving behavior, a good anti-deactivation performance, and a high hydrogenation capacity. So it will play an important role in handling the C_3 fraction from heavy cracking feedstock as having been indicated in last paragraph. As the pyrolysis process change, e.g. the replacement of raw materials by heavier oil, the MAPD content in pyrolysis gas increases while propylene decreases. Assume the contents of MAPD and propylene in C_3 fraction are about 7% and 88% respectively.

Aspen Plus simulation was performed at a hydrogen flow rate increased to 107 kg/h with a single pass conversion of 50% and a high selectivity of 99%. The MAPD content in the inlet stream of the backpack-like reactor is only 0.9%, so the reaction heat can be easily removed by the latent heat of vaporization without additional cooling unit. The bottom-draw rate was 1.13 t/h and the MAPD content was 0.55%. Propylene was drawn out from 5th tray at a rate of 20.84 t/h, where MAPD was traceless. The propylene yield has an increase of 0.9 t/h, about 4.5% of total output (Tab.1). Besides, the higher MAPD content exists in the feedstock, the more reaction heat can be used and the more energy can be saved. For pyrolysis process with heavy cracking feedstock, the energy consumed in propylene distillation is about 5.6% less than conventional process (Tab.1).

The comparison between novel propylene catalytic distillation process and conventional processes is summarized in Tab.1. The new process has a higher selectivity of MAPD conversion, a higher propylene yield and less energy consumed, and the results are much better for higher MAPD content in feedstock. In addition, the effect of deactivation on the MAPD content in the bottom-draw product is also weakened.

5. Conclusions

A novel catalytic distillation process with a high selectivity of MAPD conversion is proposed by Aspen Plus simulation. MAPD are separated from propylene in propylene fractionation column firstly and are converted into propylene by catalytic hydrogenation secondly. The selectivity of hydrogenation is improved greatly with a propylene yield increased by 2.1%~4.5%; energy is saved by 2.5%~5.6%. This process is especially suitable for heavy cracking feedstock due to a high hydrogenation capacity and a good anti-deactivation behavior.

Acknowledgements

One of the authors, Yi Liu, is grateful to Prof. G.X. Bai for helpful discussions.

Symbols

a		activity of catalyst
M		magnification of the outlet MAPD concentration after deactivation
C	[mol/L]	MAPD concentration in the feedstock
C_i	[mol/L]	inlet MAPD concentration of conventional reactor
C_o	[mol/L]	outlet MAPD concentration of conventional reactor
C_i'	[mol/L]	inlet MAPD concentration of conventional reactor after catalyst deactivation
C_o'	[mol/L]	outlet MAPD concentration of conventional reactor after catalyst deactivation
C_1	[mol/L]	inlet MAPD concentration of backpack-like reactor
C_2	[mol/L]	outlet MAPD concentration of backpack-like reactor
C_1'	[mol/L]	inlet MAPD concentration of backpack-like reactor after catalyst deactivation
C_2'	[mol/L]	outlet MAPD concentration of backpack-like reactor after catalyst deactivation
C_B	[mol/L]	MAPD concentration in the bottom-draw of distillation column
Q	[m^3/h]	volume flow rate of feedstock
Q_c	[m^3/h]	volume flow rate of recycle stream of the conventional reactor
Q_r	[m^3/h]	volume flow rate of the stream into the conventional reactor
Q_R	[m^3/h]	volume flow rate of the stream into the backpack-like reactor
Q_B	[m^3/h]	volume flow rate of bottom-draw of distillation column
k	[1/s]	kinetics constant in the conventional reactor
k_2	[1/s]	kinetics constant in the backpack-like reactor
τ	[s]	residence time in the conventional reactor
τ_2	[s]	residence time in the backpack-like reactor

References

[1] F.H. Falqi, The miracle of petrochemicals, Universial publishers, Florida, USA, 2009.

[2] S. Wang, X. He, Ethylene technology, Sinopec press, Beijing, China, 2000.

[3] M.L. Derrien, J.W. Andrews, P. Bonnifay, J. Leonard, The IFP selective hydrogenation process, Chemical Engineering Progress, 70 (1) (1974) 74-80.

[4] M.L. Derrien, Selective hydrogenation applied to the refining of petrochemical raw materials produced by steam cracking, Studies in surface science and catalysts, 27 (1986) 613-666.

[5] A.E. Krumins, H.L. Leung, L.X. Xiang, LB fining for liquid phase MAPD hydrogenation, AIChE 4th ethylene producers' conference, New Orleans, LA, USA, April 1992.

[6] E.L. Mohundro, Overview on C2 and C3 selective hydrogenation in ethylene plants, AIChE 15th Ethylene produces Conference, New Orleans, LA, USA. March, 2003.

[7] J. Ji, K. Liu, X. Zhao, Z. Yu, B. Gao, Kinetics study on liquid phase catalytic hydrogenation of C_3, Petrochemical design, 19 (1) (2002) 59-62

[8] R.T. Dickerson, Catalyst for the interconversion of propyne and propadiene, US Patent 3579600, 1971.

[9] W.K. Lam, Y.R. Mehra, D.W. Mullins, Front-end hydrogenation and absorption process for ethylene recovery, US Patent 5220097, 1993.

[10] G. Gildert, W. Groten, K. Rock, D. Weidert, Combining hydrogenation with distillation in ethylene plants, AIChE 7th ethylene producers' Conference, Houston, TX, USA, March 1995.

[11] S.J. Stanley, C. Sumner, Catalytic distillation and hydrogenation of heavy unsaturates in an olefins plant, US Patent 5925799, 1999.

[12] S.J. Stanley, G.R. Gildert, Process for the removal of MAPD from hydrocarbon streams, US Patent 6414205, 2002

[13] J.Y. Ryu, Process for the selective hydrogenation of alkynes, US Patent 7041860, 2006.

[14] A.P. Gelbein, L.A. Smith, Selective hydrogenation of acetylenes and dienes in a hydrocarbon stream, US Patent 7368617, 2006

[15] C.S. Crossland, J.R. Adams, Multi-functional catalytic distillation structure, US Patent 5942456, 1999.

[16] L.Liao, J. Cheng, Z. Wang, Q. Wu, Study on the process of C3 fraction catalytic hydrogenation with distillation and its catalyst, Journal of chemical industry & engineering, 24 (2) (2003) 18~20.

[17] Y. Yang, W. Dai, D. Li, S. Chen, J. Cheng, etc. A catalytic distillation method for olefin production, CN Patent 1274649, 2006.

[18] G.M. Perry. Hydrocarbon separation, US Patent 6066238, 2000

[19] C.F. Van Egmond, Distillation process for removal of methyl acetylene and/or propadiene from an olefin stream, US Patent 7402720, 2008.

Figure Captions

Fig.1 Conventional tail-end hydrogenation processes in fixed bed reactor(s):

(a) Sequential or front-end deethanizing separation procedure;

(b) Front-end depropanizing separation procedure.

Fig.2 Conventional single-stage hydrogenation reactor

Fig.3 Catalytic distillation processes: (a) Conventional process; (b) Novel process.

Fig.4 A novel propylene catalytic distillation process

Fig.5 Comparison between novel propylene catalytic distillation column and conventional propylene distillation column: (a) Vapor/liquid loads; (b) Temperature profiles; (c) Composition profiles of propylene and propane; (d) Composition profiles of MAPD.

Fig.6 A backpack-like propylene catalytic distillation process

Tab.1 Comparison between new and conventional processes

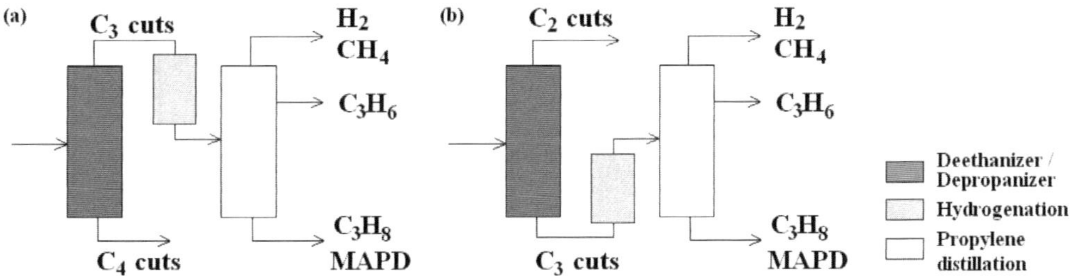

Fig.1 Conventional tail-end hydrogenation processes in fixed bed reactor(s):
(a) Sequential or front-end deethanizing separation procedure;
(b) Front-end depropanizing separation procedure.

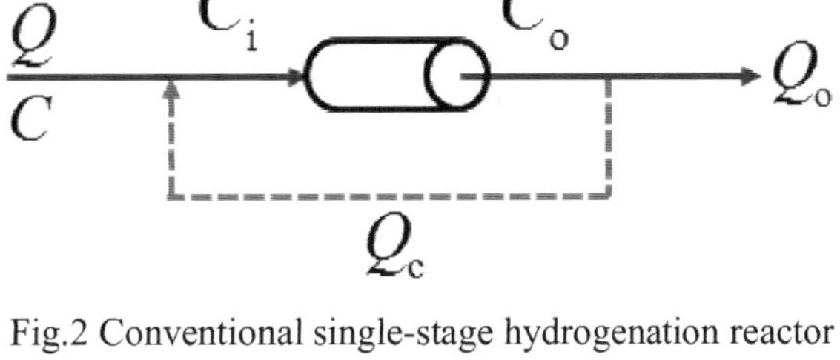

Fig.2 Conventional single-stage hydrogenation reactor

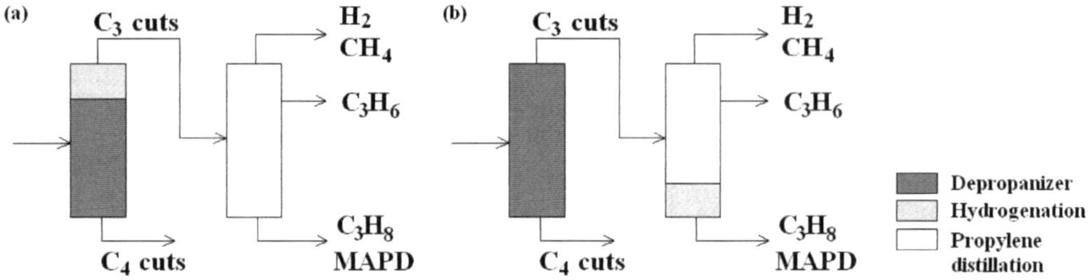

Fig.3 Catalytic distillation processes: (a) Conventional process; (b) Novel process.

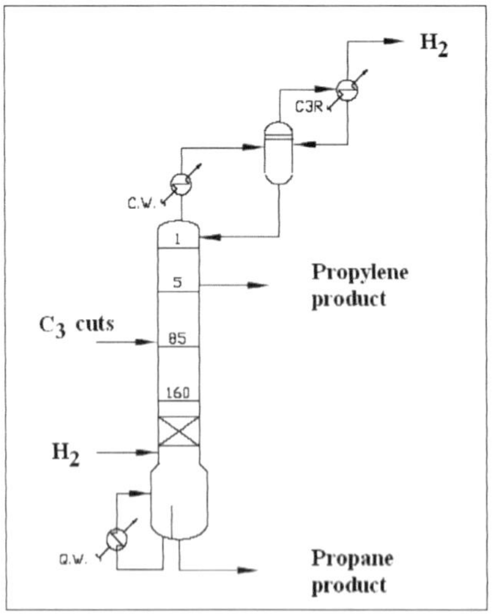

Fig.4 A novel propylene catalytic distillation process

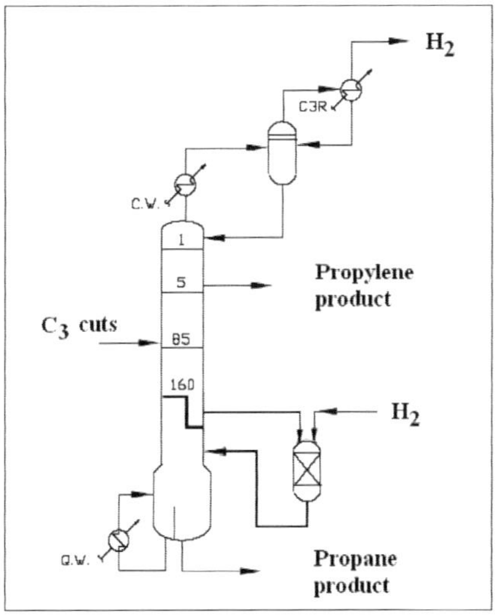

Fig.6 A backpack-like propylene catalytic distillation process

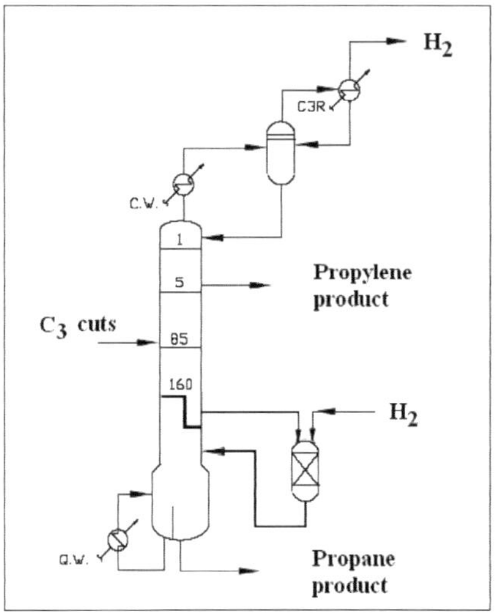

Fig.6 A backpack-like propylene catalytic distillation process

Tab.1 Comparison between new and conventional processes

Composition of feedstock	MAPD	3.5%		7.0%	
	Propylene	92.8%		88.0%	
Process		Old	New	Old	New
Outlet Composition after hydrogenation	Propylene	94.1%	2.1%	90.1%	4.3%
	MAPD	207 ppm	0.22%	209 ppm [b]	0.45% [a]
MAPD content (bottom-draw product)	100% activity	0.38%	0.27%	0.24% [b] (2.36% [a])	0.55% [a]
	50% activity	3.82% [a]	0.66% [a]	-	-
	10% activity	-	3.67% [a]	-	-
Hydrogenation selectivity		50%	99%	40%	99%
Hydrogen consumed	kg/h	79	54	168	107
Propylene yield	t/h	20.73	21.16	19.95	20.84
	increased by, t/h	-	0.43	-	0.9
	increased by, %	-	2.1	-	4.5
Energy consumed	MW	27.84	27.13	27.84	26.29
	saved by, %	-	2.5	-	5.6

[a] Reactor(s) not changed, [b] Reactor(s) need to be revised.

AUTHOR INDEX

AUTHOR INDEX